PROFESSIONAL

STANDARDS

FOR TEACHING MATHEMATICS

STANDARDS

PROFESSIONAL
STANDARDS
FOR TEACHING MATHEMATICS

Prepared by the Working Groups of the
Commission on Teaching Standards for School Mathematics
of the National Council of Teachers of Mathematics

March 1991

 NATIONAL COUNCIL OF
TEACHERS OF MATHEMATICS

The NCTM Commission on Teaching Standards for School Mathematics

Copyright © 1991 by
THE NATIONAL COUNCIL OF TEACHERS OF MATHEMATICS, INC.
1906 Association Drive, Reston, Virginia 22091-1593
All rights reserved

Fourth printing 1994

Library of Congress Cataloging-in-Publication Data:

National Council of Teachers of Mathematics. Commission on Teaching Standards for
   School Mathematics.
    Professional standards for teaching mathematics / prepared by the working groups
of the Commission on Teaching Standards for School Mathmatics of The National
Council of Teachers of Mathematics.
      p.   cm.
    "March 1991."
    Includes bibliographical references.
    ISBN 0-87353-307-0 : $25.00
    1. Mathematics—Study and teaching—Standards.   I. Title.
QA11.N29 1991   90-26154
510'.71—dc20   CIP

Printed in the United States of America

TABLE OF CONTENTS

This material is based upon work supported by the National Science
Foundation under Grant No. TPE-8954756. The Government has
certain rights in this material. Any opinions, findings, and conclusions
or recommendations expressed in this material are those of the
author(s) and do not necessarily reflect the views of the National
Science Foundation.

Photographs are by Patricia Fisher, except the photo on p. 28, which is by Michael C. George.

PREFACE

In early 1989 the Commission on Professional Teaching Standards was established by the Board of Directors of the National Council of Teachers of Mathematics. The commission was charged to produce a set of standards that promotes a vision of mathematics teaching, evaluating mathematics teaching, the professional development of mathematics teachers, and responsibilities for professional development and support, all of which would contribute to the improvement of mathematics education as envisioned in the *Curriculum and Evaluation Standards for School Mathematics.*

The standards were drafted in the summer of 1989 and revised in the summer of 1990 by the members of the commission and the three working groups, each representing a cross section of the mathematics education community, including classroom teachers, supervisors, educational researchers, mathematics teacher educators, and university mathematicians (see page ii). They were appointed by Shirley Frye, then president of NCTM.

The meetings of the commission and the working groups were held at Michigan State University. All who worked on the document thank Frank Hoppensteadt, Dean of the College of Natural Science, and Judith Lanier, Dean of the College of Education, and the many faculty members who helped to make working conditions for the group ideal. In addition we owe a debt of gratitude to Nan Jackson, Janine Remillard, and Kara Suzuka for their excellent contribution to the final drafting of the document and for their fine work in organizing and coding the hundreds of written responses to the first draft of the document.

The revisions were based on the very thoughtful responses to the working draft of this document gathered from individuals and groups during the 1989–90 school year. This document is much stronger and more coherent because of the careful reviews and suggestions that were provided. We thank all who contributed comments and hope that you see the results of your reactions in this final document. The stories of teaching that are used to craft the vignettes represent many of your suggestions. We are confident that this document represents the consensus of NCTM's members about teaching mathematics, evaluating the teaching of mathematics, the professional development of teachers, and responsibilities for the support and development of teachers and teaching.

The *Professional Standards for Teaching Mathematics* is designed, along with the *Curriculum and Evaluation Standards for School Mathematics,* to establish a broad framework to guide reform in school mathematics in the next decade. In particular, these standards present a vision of what teaching should entail to support the changes in curriculum set out in the *Curriculum and Evaluation Standards.* This document spells out what teachers need to know to teach toward new goals for mathematics education and how teaching should be evaluated for the purpose of improvement. We challenge all who have responsibility for any part of the support and development of mathematics teachers and teaching to use these standards as a basis for discussion and for making needed change so that we can reach our goal of a quality mathematics education for every child.

ACKNOWLEDGMENTS

Preparation for the NCTM standards originated more than a decade ago with the release in 1980 of *An Agenda for Action: Recommendations for School Mathematics for the 1980s*. Since then, the goals and activities of the Council have sustained and shaped the emphasis on curricular reform. The continued commitment to the evolving process of change in school mathematics has become an NCTM baton, passed on by each president to the next. This document, the companion to the *Curriculum and Evaluation Standards for School Mathematics*, represents the current step in the continuum: it advances and expands the vision of a high-quality mathematics education for every child.

The Council acknowledges with gratitude the outstanding leadership of Glenda Lappan, who chaired the Commission on Teaching Standards for School Mathematics and effectively directed the writing project. The writers and working groups deserve great credit for their scholarship, diligence, and dedication. We are also grateful for the active participation of NCTM's membership in the review process along with that of the other organizations and thousands of individuals both within and outside the profession who volunteered their reactions.

We appreciate the support and total involvement of the Headquarters staff throughout every stage of the project. Our executive director, James Gates, who guided that vital process, deserves full recognition and our thanks.

F. Joe Crosswhite, President 1984–1986

John A. Dossey, President 1986–1988

Shirley M. Frye, President 1988–1990

Iris M. Carl, President 1990–1992

INTRODUCTION

Background and Rationale

In March 1989, the National Council of Teachers of Mathematics (NCTM) released its *Curriculum and Evaluation Standards for School Mathematics*. These standards were the result of three years of planning, writing, and consensus-building among the membership of NCTM and the broader mathematics, science, engineering, and education communities, the business community, parents, and school administrators. The document describes what a high-quality mathematics education for North American students, K–12, should comprise. Central to the *Curriculum and Evaluation Standards* is the development of mathematical power for all students. Mathematical power includes the ability to explore, conjecture, and reason logically; to solve nonroutine problems; to communicate about and through mathematics; and to connect ideas within mathematics and between mathematics and other intellectual activity. Mathematical power also involves the development of personal self-confidence and a disposition to seek, evaluate, and use quantitative and spatial information in solving problems and in making decisions. Students' flexibility, perseverance, interest, curiosity, and inventiveness also affect the realization of mathematical power.

To reach the goal of developing mathematical power for all students requires the creation of a curriculum and an environment, in which teaching and learning are to occur, that are very different from much of current practice. The image of mathematics teaching needed includes elementary and secondary teachers who are more proficient in—

♦ selecting mathematical tasks to engage students' interests and intellect;

♦ providing opportunities to deepen their understanding of the mathematics being studied and its applications;

♦ orchestrating classroom discourse in ways that promote the investigation and growth of mathematical ideas;

♦ using, and helping students use, technology and other tools to pursue mathematical investigations;

♦ seeking, and helping students seek, connections to previous and developing knowledge;

♦ guiding individual, small-group, and whole-class work.

This is a considerable change from the descriptions of mathematics classes drawn from the NSF case studies (Welch 1978, p. 6):

> In all math classes that I visited, the sequence of activities was the same. First, answers were given for the previous day's assignment. The more difficult problems were worked on by the teacher or the students at the chalkboard. A brief explanation, sometimes none at all, was given of the new material, and the problems assigned for the next day. The remainder of the class was devoted to working on homework while the teacher moved around the room answering questions. The most noticeable thing about math classes was the repetition of this routine.

Even though these observations were made over ten years ago, there is little indication that the situation is different today. The routine described continues (NCTM, 1989; National Research Council 1989; Weiss 1989).

There are many persistent obstacles to making significant changes in mathematics teaching and learning in schools. Among these are the beliefs and dispositions that both students and teachers bring to the mathematics classroom, as well as the assumptions held by school administrators, parents, and society in general about mathematics curriculum and instruction. In order to change our perspective about mathematics teaching and learning, we need direction on how mathematics can be taught and learned to enhance the development of mathematical power.

In early 1989, NCTM established a commission to produce a set of *Professional Standards for Teaching Mathematics* as a companion to the *Curriculum and Evaluation Standards*. The goal of this second set of standards was to provide guidance to those involved in changing mathematics teaching. Together, these two sets of standards are a part of NCTM's long-term commitment to provide direction for the reform of school mathematics. As teachers and administrators, school districts, states, provinces, certification boards, university faculty, and other groups propose solutions to curricular, teaching, and evaluation issues in mathematics education, these two sets of standards can be used as criteria against which their ideas can be compared and judged.

Key Figures in Change

The *Professional Standards for Teaching Mathematics* rests on the following two assumptions:

♦ Teachers are key figures in changing the ways in which mathematics is taught and learned in schools.
♦ Such changes require that teachers have long-term support and adequate resources.

Educational research findings from cognitive psychology and mathematics education indicate that learning occurs as students actively assimilate new information and experiences and construct their own meanings (Case and Bereiter 1984; Cobb and Steffe 1983; Davis 1984; Hiebert 1986; Lampert 1986; Lesh and Landau 1983; Schoenfeld 1987). This is a major shift from learning mathematics as accumulating facts and procedures to learning mathematics as an integrated set of intellectual tools for making sense of mathematical situations (Resnick 1987). This view of learning is summarized in *Everybody Counts* (National Research Council 1989, pp. 58–59):

> Effective teachers are those who can stimulate students to *learn* mathematics. Educational research offers compelling evidence that students learn mathematics well only when they *construct* their own mathematical understanding. To understand what they learn, they must enact for themselves verbs that permeate the mathematics curriculum: "examine," "represent," "transform," "solve," "apply," "prove," "communicate." This happens most readily when students work in groups, engage in discussion, make presentations, and in other ways take charge of their own learning.

All students engage in a great deal of invention as they learn mathematics; they impose their own interpretation on what is presented to create a theory that makes sense to them. Students do not learn simply a subset of what they have been shown. Instead, they use new information to modify their prior beliefs. As a consequence, each student's knowledge of mathematics is uniquely personal.

The kind of teaching envisioned in these standards is significantly different from what many teachers themselves have experienced as students in mathematics classes. Because teachers need time to learn and develop this kind of teaching practice, appropriate and ongoing professional development is crucial. Good instructional and assessment materials and the latitude to use them flexibly are also keys to the process of change.

For teachers to be able to change their role and the nature of their classroom environment, administrators, supervisors, and parents must expect, encourage, support, and reward the kind of teaching described in this set of standards. We cannot expect teachers to respond simultaneously to several different calls for change or other new demands. Change is difficult and will take time and reliable, systematic support.

Major Shifts

Woven into the fabric of the *Professional Standards for Teaching Mathematics* are five major shifts in the environment of mathematics classrooms that are needed to move from current practice to mathematics teaching for the empowerment of students. We need to shift—

- toward classrooms as mathematical communities—away from classrooms as simply a collection of individuals;
- toward logic and mathematical evidence as verification—away from the teacher as the sole authority for right answers;
- toward mathematical reasoning—away from merely memorizing procedures;
- toward conjecturing, inventing, and problem solving—away from an emphasis on mechanistic answer-finding;
- toward connecting mathematics, its ideas, and its applications—away from treating mathematics as a body of isolated concepts and procedures.

As teachers shift toward the vision of teaching presented by these standards, one would expect to see teachers asking, and stimulating students to ask, questions like the following:

- ### Helping students work together to make sense of mathematics

    "What do others think about what Janine said?"

    "Do you agree? Disagree?"

    "Does anyone have the same answer but a different way to explain it?"

    "Would you ask the rest of the class that question?"

    "Do you understand what they are saying?"

    "Can you convince the rest of us that that makes sense?"

- ### Helping students to rely more on themselves to determine whether something is mathematically correct

    "Why do you think that?"

    "Why is that true?"

    "How did you reach that conclusion?"

    "Does that make sense?"

    "Can you make a model to show that?"

- ### Helping students learn to reason mathematically

    "Does that always work?"

    "Is that true for all cases?"

    "Can you think of a counterexample?"

"How could you prove that?"

"What assumptions are you making?"

♦ *Helping students learn to conjecture, invent, and solve problems*

"What would happen if . . .? What if not?"

"Do you see a pattern?"

"What are some possibilities here?"

"Can you predict the next one? What about the last one?"

"How did you think about the problem?"

"What decision do you think he should make?"

"What is alike and what is different about your method of solution and hers?"

♦ *Helping students to connect mathematics, its ideas, and its applications*

"How does this relate to... ?"

"What ideas that we have learned before were useful in solving this problem?"

"Have we ever solved a problem like this one before?"

"What uses of mathematics did you find in the newspaper last night?"

"Can you give me an example of...?"

All Students

Throughout these standards the phrase *all students* is used often. By this phrase we mean to set the mathematical education of every child as the goal for mathematics teaching at all levels, K–12. In April 1990, the NCTM Board of Directors endorsed the following statement:

> As a professional organization and as individuals within that organization, the Board of Directors sees the comprehensive mathematics education of every child as its most compelling goal.
>
> By "every child" we mean specifically—
>
> ♦ students who have been denied access in any way to educational opportunities as well as those who have not
> ♦ students who are African American, Hispanic, American Indian, and other minorities as well as those who are considered to be a part of the majority;
> ♦ students who are female as well as those who are male; and
> ♦ students who have not been successful in school and in mathematics as well as those who have been successful.

It is essential that schools and communities accept the goal of mathematical education for every child. However, this does not mean that every child will have the same interests or capabilities in mathematics. It does mean that we will have to examine our fundamental expectations about what children can learn and can do and that we will have to strive to create learning environments in which raised expectations for children can be met.

Components of the Professional Teaching Standards

The kind of instruction needed to implement the NCTM *Standards* requires a high degree of individual responsibility, authority, and autonomy—in short, professionalism on the part of each teacher. To give guidance to the development of such professionalism in mathematics teaching, the *Professional Standards for Teaching Mathematics* consists of five components:

1. Standards for teaching mathematics
2. Standards for the evaluation of the teaching of mathematics
3. Standards for the professional development of teachers of mathematics
4. Standards for the support and development of mathematics teachers and teaching
5. Next steps

Standards for Teaching Mathematics. This section develops a vision of what a teacher at any level of schooling must know and be able to do to teach mathematics as envisioned by the NCTM *Curriculum and Evaluation Standards for School Mathematics* and the *Professional Standards for Teaching Mathematics*. The standards in this section are organized around a framework emphasizing the important decisions that a teacher makes in teaching—

- Setting goals and selecting or creating mathematical *tasks* to help students achieve these goals;
- Stimulating and managing classroom *discourse* so that both the students and the teacher are clearer about what is being learned;
- Creating a classroom *environment* to support teaching and learning mathematics;
- *Analyzing* student learning, the mathematical tasks, and the environment in order to make ongoing instructional decisions.

The statements of standards focus on the major aspects of good mathematics teaching across all grade levels. Guides for teaching at different levels are suggested in the elaborations of each standard and in the annotated vignettes that are used to provide examples. These vignettes show a range of situations in which good mathematics teaching and learning can take place. A high-quality mathematics experience is not determined simply by the presence of computers or calculators or the use of small groups, manipulatives, or student discussions. The nature of the mathematical task posed and what is expected of students are critical aspects against which to judge the effectiveness of the lesson. Although each vignette could illustrate many of the standards, the guiding comments focus on the particular standard being elaborated.

Standards for the Evaluation of Teaching of Mathematics. This section presents NCTM's vision for the evaluation of mathematics teaching. The assumption is that the major purpose for such evaluation is the improvement of teaching. The standards include the following aspects of the evaluation of teaching:

- The process of evaluation
- The foci of evaluation

In March 1987, the NCTM Board of Directors approved a position statement on the Evaluation of Teacher Performance, which includes the following:

Evaluation includes the identification of goals by the teacher and the evaluators, the collection of information, and a collaborative dialogue between the teacher and evaluators to reformulate, redirect, and refine goals for the future. Goals for personal and professional growth may include some that are mandated by the state or province, district, teacher education institution, or individual school, but the teacher must be an active participant in identifying goals of a more specific nature.

The use that is made of the information gained through the evaluation process is as important as the act of evaluation itself. The appropriate outcome of this ongoing process is a collaborative dialogue between the teacher and others involved in the process, resulting in a mutually agreed-on plan for professional growth.

The standards in this section are consistent with this position statement on evaluation and support the assumption that the evaluation of teaching should result in the professional growth of teachers. These standards give guidance to teachers seeking self-improvement, to colleagues mentoring others, and to supervisors and others who are involved in the evaluation of teaching.

The vignettes in this section show a range of assessment activities and personnel involved in evaluation. They illustrate the substance and process of assessing teaching and of the results of evaluation and are not meant to be exhaustive of the possibilities.

Standards for the Professional Development of Teachers of Mathematics. This section expresses NCTM's vision for well-prepared teachers of mathematics from the time prospective teachers of mathematics take their first courses in collegiate mathematics throughout their career-long development. These standards focus on what a teacher needs to know about mathematics, mathematics education, and pedagogy to be able to carry out the vision of teaching discussed in the first component of this document. The following aspects of both the preservice and in-service phases of the professional development of teachers are addressed:

♦ Modeling good mathematics teaching

♦ Knowing mathematics and school mathematics

♦ Knowing students as learners of mathematics

♦ Knowing mathematical pedagogy

♦ Developing as a teacher of mathematics

♦ Teachers' roles in professional development

These teaching standards provide essential guidance to colleges, universities, and schools; state departments and provincial ministries of education; public and private schools; and all who are a part of the preparation and professional development of teachers. These standards focus attention on the roles of faculty in college and university departments of education and mathematics and school officials responsible for professional development. They also emphasize the need for dialogue among these partners in nurturing excellence in mathematics teaching.

The current reform movement in mathematics education, and in education in general, has as a strong underlying theme the professionalism of teaching. This view recognizes the teacher as a part of a learning community that continually fosters growth in knowledge, stature, and responsibility. The standards in this section provide a guide to the preparation, support, and career development of teachers. The standards themselves are meant to be general principles that can be used to improve the quality of teacher preparation programs as well as school and university professional-development activities. Applications of these general principles to levels of preparation and phases of career development are illustrated in the elaborations and the vignettes.

Standards for the Support and Development of Teachers and Teaching. The standards in this section spell out the responsibilities of those who make decisions that affect teaching mathematics. The responsibilities of the following groups are addressed:

♦ Policy makers in government, business, and industry

♦ Schools and school systems

♦ Colleges and universities

♦ Professional organizations

Decisions made by others can enable teachers to move toward the vision of teaching described in these standards or can constrain the mathematics program in ways that cripple efforts to improve teaching. The environment in which teachers teach is as important to their success as the environment in which students learn is to theirs. These standards highlight responsibilities and ways in which others can support teachers in shifting toward the vision of teaching needed to support the implementation of the *Curriculum and Evaluation Standards for School Mathematics.*

Next Steps. The final section of this document discusses some of the issues and next steps that we can take to move toward our goal of mathematical power for all students.

Conclusions

These teaching standards are not intended to be an exhaustive checklist of specific concepts, skills, and behaviors that teachers must have. Instead, these standards are a set of principles accompanied by illustrations or indicators that can be used to judge what is valuable and appropriate. They give direction for moving toward excellence in teaching mathematics. They furnish guidance to all who are interested in improving teaching, including teachers, universities, state departments of education and provincial ministries of education, local school districts, private schools, teacher organizations, the National Council for Accreditation of Teacher Education, the National Board on Professional Teaching Standards, and others who license or certify teachers or who evaluate teaching or teacher education programs.

FIRST STEPS

This account of a sixth-grade teacher introduces the reader to the *annotated vignettes* that are used throughout this document to elaborate the visions of teaching, the evaluation of teaching, and professional development. Narratives—drawn from actual school and university classrooms with a range of teachers and students in a variety of contexts—are annotated throughout in italics. The narratives are meant to be like video clips. They provide brief but vivid glimpses into diverse settings and help to build depth into the images created by this document. As such, they are intended to animate the standards: they illustrate points discussed in the text and make the issues multidimensional. Although the vignettes do exemplify some specific worthwhile practices, they do not suggest one "correct" approach to teaching mathematics. In the following introductory vignette, the comments on the right foreshadow issues that are examined throughout this volume. These annotations are intended to help orient the reader to the sections that follow.

Vignette

Three days remain until the beginning of school. Sharon Robinson is sitting in her classroom, leafing through materials from the summer school class she took as part of her master's program at the nearby college. She really liked the course. It included a stimulating mix of new ideas, opportunities to experiment, and time for the teachers enrolled in the course to think and talk with one another. Former classroom teachers themselves, the two professors who team-taught the course seemed sensible and realistic, and yet they clearly had a different orientation to mathematics teaching.

The Professional Development section discusses teachers' responsibilities for their own professional development and takes the position that good mathematics teaching should be modeled in teachers' professional development experiences.

The eight-week course had been exactly what Sharon needed, for she had finished the school year in June feeling vaguely dissatisfied with her mathematics teaching. After five years of teaching, she had become able to manage her classroom effectively, to cover the required curriculum, and to incorporate some neat supplementary activities. Sharon received an excellent evaluation from her principal, Mrs. Bowdoin. The principal, however, had little mathematics background and her comments were always focused on management issues. These did not address Sharon's growing questions about her mathematics teaching.

Both the Evaluation and the Teaching sections describe ways in which teachers can analyze their own teaching.

The Evaluation section addresses the need for teachers to receive support from mathematics specialists.

Her sixth graders typically did quite well on the district mathematics test. Still, Sharon was troubled about her students' participation in, and success with, mathematical reasoning and problem solving. For example, she noticed that her boys talked and were much more active than her girls. She is also aware—and concerned—that many of her African American and Hispanic students did not go on to take algebra in the eighth or ninth grade. She wants to do something to affect these patterns of participation. In general, she felt that her students lacked both confidence and skills with mathematical reasoning and problem solving. If she gave her students word problems to do, for example, they often gave up easily or came up with answers that made no sense at all.

The Teaching section deals with orchestrating the discourse in the learning environment. Both the Teaching and the Evaluation sections stress the commitment to helping EVERY student develop mathematical power.

In the summer class, the teachers became familiar with the *Curriculum and Evaluation Standards*. The professors also planned sessions in which Sharon and the other teachers had opportunities to engage in mathematical activity in ways that they had never experienced before. Class members found themselves—to their surprise—wrapped up in the problems, excited about trying to convince one another of their solutions, and genuinely interested in alternative pathways and approaches.

The Professional Development section deals with the need for teachers themselves to be engaged in interesting mathematics and in mathematical discourse as a part of their professional growth.

The Teaching section elaborates on the dimensions of teaching—tasks, discourse, environment—that are involved in changing one's approach; it makes plain that this document offers a vision, not a recipe, for creating new practices.

Evaluation should support and enhance a teacher's professional growth. Both the Evaluation and the Support sections discuss the role administrators play.

How teachers can support one another's professional growth is a continuing theme in this volume. All sections stress the importance of teachers paying attention to students' knowledge and their ways of thinking about mathematics.

The Teaching section elaborates criteria for fruitful mathematical tasks.

Teachers who strive to change their mathematics teaching in the directions outlined by these standards are in the position of creating and reflecting on new practices.

Now, with the start of school just a few days away, Sharon is determined to begin forging a new approach in her own teaching. She realizes that what she has in mind is quite different. It involves not just new techniques but a new way of thinking about mathematics and mathematics teaching. Suddenly she worries: Will Mrs. Bowdoin understand that kids may not be sitting so quietly and listening just to me so much this year? Does she realize that they will be—I hope—talking more, giving reasons for their answers, coming up with new ideas and questions? Sharon hopes that she can explain to her principal that she is going to be experimenting a bit as she tries to change the way she teaches mathematics but that she is not sure what form this will take as the year unfolds. She hopes that Mrs. Bowdoin will support her efforts and take them into account when she evaluates her this year.

Tom Flood, Sharon's colleague, wanders into the room, and she describes her quandary to him—how to start the new year? She wants to begin to shape the classroom in this new direction. She wants to promote more conjecturing and problem solving. She also wants to learn more about her students—how they think, what they know and can do, how they feel about mathematics. On the basis of these concerns, Sharon is considering using the following problem:

Take 12 tiles and build rectangles with them. How many rectangles can you make using all 12 tiles?

Now try 16. How many can you make using all 16?

Now find another number of tiles that will let you make MORE rectangles than you can make with 12 tiles.

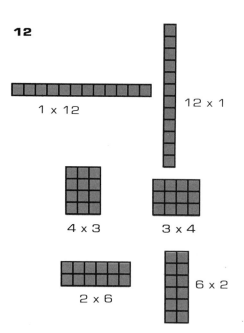

She explains that she likes this problem because it also gives students an opportunity to make some nice connections—between number theory and geometry, for instance.

"It *is* a nice problem," nods Tom. "You will be able to learn a lot about what they already know—like about factors, about what a rectangle is, about how to work on mathematics problems...."

"Yes!" Sharon breaks in. "When we did a problem like this in my summer school class, an important part of it was the question of how you knew when you had all the possible rectangles and the idea of *proving* that to the other people in the group. I had never thought about that as an important issue at all. Like with 12 tiles, you know you are done when you have built a 1 × 12, a 2 × 6, a 3 × 4—and the opposites of those, like 12 × 1, 6 × 2, and so on. You can prove that you have finished because you have taken all the numbers that divide 12—all its factors— and made rectangles with them." The words are tumbling out of Sharon's mouth. "And I thought—just as you said—that I could learn a lot about what they know and about their dispositions toward mathematics. I wanted to do more than just set the tone for the year, although that is a good idea, too. I hope I manage that."

"But why did you choose 12—and then why 16?" asks Tom.

"Well, 12 makes six rectangles, but 16 makes only five. I think that will surprise them. And I'm not sure they will consider the 4 × 4 shape a rectangle. I bet they think a square is not a rectangle. I would like to get them to examine that," explains Sharon.

Tom thinks about this for a moment. "So is that why you ask about 16 tiles before you ask the last question—about finding one that will make more rectangles than 12? Aren't you the clever one!"

Sharon laughs. The two discuss the activity further and find themselves having fun considering ways to extend the problem or have the students generate extensions—some to pursue perhaps now and others to return to later in the year. They keep a list:

♦ What other numbers of tiles will make an odd number of rectangles? Why is that?

Teachers need opportunities to engage as learners in well-taught mathematics courses and workshops. Aspects of this idea are discussed in the Professional Development section.

Knowledge of students' understandings and ways of thinking helps teachers to construct worthwhile mathematical tasks. This is explored in the Teaching section.

Encouraging students to formulate problems on their own is an aspect of problem solving that is emphasized in both the Teaching and the Evaluation sections.

[squares!]

1	1 × 1
2	1 × 2, 2 × 1
3	1 × 3, 3 × 1
4	1 × 4, 4 × 1, 2 × 2
5	1 × 5, 5 × 1
6	1 × 6, 6 × 1, 2 × 3, 3 × 2
7	1 × 7, 7 × 1
8	1 × 8, 8 × 1, 2 × 4, 4 × 2
9	1 × 9, 9 × 1, 3 × 3
.	.
.	.
.	.
16	1 × 16, 16 × 1, 2 × 8, 8 × 2, 4 × 4

♦ ♦ ♦ ♦ ♦ ♦ ♦ ♦ ♦

♦ What numbers of tiles will make the fewest rectangles? Why is that?

[primes]

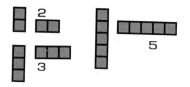

♦ What numbers of tiles can produce exactly three rectangles? Or six? Is there a pattern?

4	$1 \times 4, 4 \times 1, 2 \times 2$
9	$1 \times 9, 9 \times 1, 3 \times 3$
25	$1 \times 25, 25 \times 1, 5 \times 5$
49	$1 \times 49, 49 \times 1, 7 \times 7$
.	.
.	.
.	.

Sharon and Paul continue to brainstorm, and they come up with still more things to explore later. They consider the possibility of exploring, in some related ways, triangular arrays, or pursuing similar investigations in three dimensions.

Tom has to meet with a parent. Before he leaves, he indicates that he will also try the same problem with his class and suggests that they compare notes afterward. He admits that he, too, has been thinking about making some changes in the ways he approaches mathematics in his class.

Sharon jots down a few more notes from their work together. Her mind turns to the problem of how she can get her sixth graders used to a different pattern of discourse, one in which answers will be determined to be right by whether they make sense and can be *proved* or *explained*—not by whether she says, "Good job!" She realizes that her students are used to the teacher being the one who tells you if you're right—that when other kids are talking, it has nothing to do with you.

Sharon wonders how her students will respond to her asking them, "What do you all think about what So-and-so just said?" She remembers how one of the teachers in the summer workshop told how when she first asked questions like that, kids just stared at her, quite confused. She wonders whether they will even be *able* to explain their answers.

Then, her mind shifting, Sharon starts thinking about what she is going to explain to parents about the mathematics program. She knows that they will be expecting their kids to be given a placement test and then started in textbooks on the basis of their performance on those tests. Sharon is thinking about working with the class more as a whole group than she has in the past. She realizes, too, that she will be having the kids do fewer traditional worksheets and so the parents will be getting less standard written computational work sent home. The written work will probably look pretty different from previous years, too. Last year, for example, on the first day of school, she gave a sheet with seventy-two mixed computational exercises—to see where the students were. This year, she smiles, she is still going to find out a lot—probably *more*—about what they know, but her strategy—using the tile problem—is very different.

Sharon decides that, in addition to writing a letter to parents about what she is trying to do (in which she will refer to the *Curriculum and Evaluation Standards*), she will have a parent meeting about mathematics—maybe in October. She considers the possibility of demonstrating and explaining to the parents some of the mathematical reasoning and problem-solving activities their kids are doing. She realizes that many of her parents work the night shift at the local plant, and so she decides to do a breakfast meeting one day. Maybe she'll even invite them to stay and watch a mathematics lesson afterward. That would give them a feel for the nature of the mathematics class—and if she can think of a good example, it would be great if she could get the parents hooked into the problem, too. This way, maybe she could demonstrate the kind of understanding and discourse she is trying to foster and how she is going about it.

Both teachers and school administrators have responsibilities to work with the community and with parents, educating them about new goals and practices in mathematics teaching. Working with parents and in the community is crucial to making change possible. Both the section on Support and the Teaching section deal with this idea.

Sharon sits back with a sigh and stretches. There is a lot to do. Still, she is excited at the prospects that lie ahead. And she is glad that Tom wants to work on this too—it will really help to have someone to talk to. But she needs to take a break from all this thinking. With one last gulp of strong lukewarm coffee, she rises and returns to the task of putting up her bulletin boards.

The importance of teachers working together and learning from one another is emphasized throughout all sections.

STANDARDS FOR TEACHING MATHEMATICS

OVERVIEW

This section presents six standards for the teaching of mathematics organized under four categories.

Tasks

1. Worthwhile Mathematical Tasks

Discourse

2. Teacher's Role in Discourse

3. Students' Role in Discourse

4. Tools for Enhancing Discourse

Environment

5. Learning Environment

Analysis

6. Analysis of Teaching and Learning

INTRODUCTION

The *Curriculum and Evaluation Standards for School Mathematics* represents NCTM's vision of what students should learn in mathematics classrooms. Congruent with the aims and rhetoric of the current reform movement in mathematics education (e.g., National Research Council 1989, 1990), the *Standards* is threaded with a commitment to developing the mathematical literacy and power of all students. Being mathematically literate includes having an appreciation of the value and beauty of mathematics as well as being able and inclined to appraise and use quantitative information. Mathematical power encompasses the ability to "explore, conjecture, and reason logically, as well as the ability to use a variety of mathematical methods effectively to solve nonroutine problems" and the self-confidence and disposition to do so (National Council of Teachers of Mathematics 1989, p. 5). It also includes being able to formulate and solve problems, to judge the role of mathematical reasoning in a real-life situation, and to communicate mathematically.

The vision of the *Curriculum and Evaluation Standards* is that mathematical reasoning, problem solving, communication, and connections must be central. Computational algorithms, the manipulation of expressions, and paper-and-pencil drill must no longer dominate school mathematics. Beyond the standard fare of number concepts and operations, the school curriculum must include serious exploration of geometry, measurement, statistics, probability, algebra, and functions. Whether working individually or in small or large groups, students should encounter, develop, and use mathematical ideas and skills in the context of genuine problems and situations. In so doing, they should develop the ability to use a variety of resources and tools, such as calculators and computers and concrete, pictorial, and metaphorical models. They must know and be able to choose appropriate methods of computation, including estimation, mental calculation, and the use of technology. As they explore and solve problems, they must engage in conjecture and argument.

In setting these goals for school mathematics, the *Curriculum and Evaluation Standards* implies a significant departure from the traditional practices of mathematics teaching. It suggests changes in not only *what* is taught but also *how* it is taught. Teachers and students have different roles in such classrooms and different notions about what it means to know and to do mathematics. The purpose of this section is to make explicit and expand the images of teaching and learning implicit in the *Curriculum and Evaluation Standards for School Mathematics*, to elaborate a vision of instruction that can light the path toward such change.

Six standards encompass the vision's core dimensions. These standards are grouped under four headings: tasks, discourse, environment, and analysis—major arenas of teachers' work that are logically central to shaping what goes on in mathematics classes.

- *Tasks* are the projects, questions, problems, constructions, applications, and exercises in which students engage. They provide the intellectual contexts for students' mathematical development.

- *Discourse* refers to the ways of representing, thinking, talking, and agreeing and disagreeing that teachers and students use to engage in those tasks. The discourse embeds fundamental values about knowledge and authority. Its nature is reflected in what makes an answer right and what counts as legitimate mathematical activity, argument, and thinking. Teachers, through the ways in which they orchestrate discourse, convey messages about whose knowledge and ways of thinking and knowing are valued, who is considered able to contribute, and who has status in the group.

- *Environment* represents the setting for learning. It is the unique interplay of intellectual, social, and physical characteristics that shapes the ways of knowing and working that are encouraged and expected in the classroom. It is the context in which the tasks and discourse are embedded; it also refers to the use of materials and space.

- *Analysis* is the systematic reflection in which teachers engage. It entails the ongoing monitoring of classroom life—how well the tasks, discourse, and environment foster the development of every student's mathematical literacy and power. Through this process, teachers examine relationships between what they and their students are doing and what students are learning.

In deciding how to present and elaborate the ideas underlying each of the six standards, we confronted two basic dilemmas. First, teaching is an integrated activity. Although we can analyze the practice of teaching into these four arenas of teachers' work—tasks, discourse, environment, and analysis—they are in fact interwoven and interdependent. The quality of the classroom environment, for example, is both a function of and an influence on the classroom discourse. Alternatively, tasks are shaped by the discourse that surrounds them and the environment in which that work takes place. Our second dilemma was that professional standards for mathematics teaching should represent values about what contributes to good practice without prescribing it. Such standards should offer a vision, not a recipe.

The format of this section grew out of consideration of these issues. Because teaching is an integrated activity and because we wanted to provide concrete images of a vision, we have chosen to use illustrative, annotated vignettes of classroom teaching and learning. The statement of each of the six standards is first elaborated with an explanation of its main ideas. Each explanation is then followed with illuminating cases that show these ideas embedded in actual teaching contexts. Drawn from transcripts, observations, and experiences in a wide variety of real classrooms, the vignettes were selected to illustrate a range of teaching styles, classroom contexts, mathematical topics, and grade levels. These vignettes were gathered from actual classrooms in a wide variety of settings, with students of diverse cultural, linguistic, and socioeconomic backgrounds. We included examples of teachers facing problems as well as cases of accomplished practice. The italicized commentaries focus on issues pertinent to that standard only, although many other features of the vision of teaching are apparent in the descriptions. For example, the vignettes in the Environment section are annotated from the perspective of the learning environment only.

ASSUMPTIONS

The standards for teaching are based on four assumptions about the practice of mathematics teaching:

1. *The goal of teaching mathematics is to help all students develop mathematical power.* The *Curriculum and Evaluation Standards for School Mathematics* furnishes the basis for a curriculum in which mathematical reasoning, communication, problem solving, and connections are central. Teachers must help every student develop conceptual and procedural understandings of number, operations, geometry, measurement, statistics, probability, functions, and algebra and the connections among ideas. They must engage all students in formulating and solving a wide variety of problems, making conjectures and constructing arguments, validating solutions, and evaluating the reasonableness of mathematical claims. Along with all this, teachers must foster in students the disposition to use and engage in mathematics, an appreciation of its beauty and utility, and a tolerance for getting stuck or sidetracked. Teachers must help students realize that mathematical thinking involves dead ends and detours and encourage them to persevere when confronted with a puzzling problem and to develop the self-confidence and interest to do so.

2. *WHAT students learn is fundamentally connected with HOW they learn it.* Students' opportunities to learn mathematics are a function of the setting and the kinds of tasks and discourse in which they participate. What students learn—about particular concepts and procedures as well as about thinking mathematically—depends on the ways in which they engage in mathematical activity in their classrooms. Their dispositions toward mathematics are also shaped by such experiences. Consequently, the goal of developing students' mathematical power requires careful attention to pedagogy as well as to curriculum.

3. *All students can learn to think mathematically.* The goals described in the *Curriculum and Evaluation Standards for School Mathematics* are goals for all students. Goals such as learning to make conjectures, to argue about mathematics using mathematical evidence, to formulate and solve problems—even perplexing ones—and to make sense of mathematical ideas are not just for some group thought to be "bright" or "mathematically able." Every student can—and should—learn to reason and solve problems, to make connections across a rich web of topics and experiences, and to communicate mathematical ideas. By "every student" we mean specifically—

- students who have been denied access in any way to educational opportunities as well as those who have not;

- students who are African American, Hispanic, American Indian, and other minorities as well as those who are considered to be part of the majority;
- students who are female as well as those who are male;
- students who have not been successful as well as those who have been successful in school and in mathematics.

This assumption is supported by the vignettes, which were drawn from classrooms with students of diverse cultural, linguistic, and socioeconomic backgrounds.

4. *Teaching is a complex practice and hence not reducible to recipes or prescriptions.* First of all, teaching mathematics draws on knowledge from several domains: knowledge of mathematics, of diverse learners, of how students learn mathematics, of the contexts of classroom, school, and society. Such knowledge is general. However, teachers must also consider the particular, for teaching is context-specific. Theoretical knowledge about adolescent development, for instance, can only partly inform a decision about particular students learning a particular mathematical concept in a given context. Second, as teachers weave together knowledge from these different domains to decide how to respond to a student's question, how to represent a particular mathematical idea, how long to pursue the discussion of a problem, or what task to use to engage students in a new topic, they often find themselves having to balance multiple goals and considerations. Making such decisions depends on a variety of factors that cannot be determined in the abstract or governed by rules of thumb.

Because teaching mathematics well is a complex endeavor, it cannot be reduced to a recipe for helping students learn. Instead, good teaching depends on a host of considerations and understandings. Good teaching demands that teachers reason about pedagogy in professionally defensible ways within the particular contexts of their own work. The standards for teaching mathematics are designed to help guide the processes of such reasoning, highlighting issues that are crucial in creating the kind of teaching practice that supports the learning goals of the *Curriculum and Evaluation Standards for School Mathematics*. This section circumscribes themes and values but does not—indeed, it could not—prescribe "right" practice.

TASKS

The mathematics tasks in which students engage—projects, problems, constructions, applications, exercises, and so on—and the materials with which they work frame and focus students' opportunities for learning mathematics in school. Tasks provide the stimulus for students to think about particular concepts and procedures, their connections with other mathematical ideas, and their applications to real-world contexts. Good tasks can help students to develop skills in the context of their usefulness. Tasks also convey messages about what mathematics is and what doing mathematics entails. Tasks that require students to reason and to communicate mathematically are more likely to promote their ability to solve problems and to make connections. Such tasks can illuminate mathematics as an intriguing and worthwhile domain of inquiry. A central responsibility of teachers is to select and develop worthwhile tasks and materials that create opportunities for students to develop these kinds of mathematical understandings, competence, interests, and dispositions.

◆　　◆　　◆　　◆　　◆　　◆　　◆　　◆

STANDARD 1:
WORTHWHILE MATHEMATICAL TASKS

The teacher of mathematics should pose tasks that are based on—

◆ *sound and significant mathematics;*

◆ *knowledge of students' understandings, interests, and experiences;*

◆ *knowledge of the range of ways that diverse students learn mathematics;*

and that

◆ *engage students' intellect;*

◆ *develop students' mathematical understandings and skills;*

◆ *stimulate students to make connections and develop a coherent framework for mathematical ideas;*

◆ *call for problem formulation, problem solving, and mathematical reasoning;*

◆ *promote communication about mathematics;*

◆ *represent mathematics as an ongoing human activity;*

◆ *display sensitivity to, and draw on, students' diverse background experiences and dispositions;*

◆ *promote the development of all students' dispositions to do mathematics.*

Elaboration

Teachers are responsible for the quality of the mathematical tasks in which students engage. A wide range of materials exists for teaching mathematics: problem booklets, computer software, practice sheets, puzzles, manipulative materials, calculators, textbooks, and so on. These materials contain tasks from which teachers can choose. Also, teachers often create their own tasks for students: projects, problems, worksheets, and the like. Some tasks grow out of students' conjectures or questions. Teachers should choose and develop tasks that are likely to promote the development of students' understandings of concepts and procedures in a way that also fosters their ability to solve problems and to reason and communicate mathematically. Good tasks are ones that do not separate mathematical thinking from mathematical concepts or skills, that capture students' curiosity, and that invite them to speculate and to pursue their hunches. Many such tasks can be approached in more than one interesting and legitimate way; some have more than one reasonable solution. These tasks, consequently, facilitate significant classroom discourse, for they require that students reason about different strategies and outcomes, weigh the pros and cons of alternatives, and pursue particular paths.

In selecting, adapting, or generating mathematical tasks, teachers must base their decisions on three areas of concern: the mathematical content, the students, and the ways in which students learn mathematics.

In considering the mathematical content of a task, teachers should consider how appropriately the task represents the concepts and procedures entailed. For example, if students are to gather, summarize, and interpret data, are the statistics they are expected to generate appropriate? Does it make sense to calculate a mean? If there is an explanation of a procedure, such as calculating a mean, does that explanation focus on the underlying concepts or is it merely mechanical? Teachers must also use a curricular perspective, considering the potential of a task to help students progress in their cumulative understanding of a particular domain and to make connections among ideas they have studied in the past and those they will encounter in the future.

A second content consideration is to assess what the task conveys about what is entailed in doing mathematics. Some tasks, although they deal nicely with the concepts and procedures, involve students in simply producing right answers. Others require students to speculate, to pursue alternatives, to face decisions about whether or not their approaches are valid. For example, one task might require students to find means, medians, and modes for given sets of data. Another might require them to decide whether to calculate means, medians, or modes as the best measures of central tendency, given particular sets of data and particular claims they would like to make about the data, then to calculate those statistics, and finally to explain and defend their decisions. Like the first task, the second would offer students the opportunity to practice finding means, medians, and modes. Only the second, however, conveys the important point that summarizing data involves decisions related to the data and the purposes for which the analysis is being used. Tasks should foster students' sense that mathematics is a changing and evolving domain, one in which ideas grow and develop over time and to which many cultural groups have contributed. Drawing on the history of mathematics can help teachers to portray this idea: exploring alternative numeration systems or investigating non-Euclidean geometries, for example. Fractions evolved out of the Egyptians' attempts to divide quantities—four things shared among ten people. This fact could provide the explicit basis for a teacher's approach to introducing fractions.

A third content consideration centers on the development of appropriate skill and automaticity. Teachers must assess the extent to which skills play a role in the context of particular mathematical topics. A goal is to create contexts that foster skill development even as students engage in problem solving and reasoning. For example, elementary school students should develop rapid facility with addition and multiplication combinations. Rolling pairs of dice as part of an investigation of probability can simultaneously provide students with practice with addition. Trying to figure out how many ways 36 desks can be arranged in equal-sized groups—and whether there are more or fewer possible groupings with 36, 37, 38, 39, or 40 desks—presses students to produce each number's factors quickly. As they work on this problem, students have concurrent opportunities to practice multiplication facts and to develop a sense of what factors are. Further, the problem may provoke interesting questions: How many factors does a number have? Do larger numbers necessarily have more factors? Is there a number that has more factors than 36? Even as students pursue such questions, they practice and use multiplication facts, for skill plays a role in problem solving at all levels. Teachers of algebra and geometry must similarly consider which skills are essential and why and seek ways to develop essential skills in the contexts in which they matter. What do students need to memorize? How can that be facilitated?

The content is unquestionably a crucial consideration in appraising the value of a particular task. Defensible reasoning about the mathematics of a task must be based on a thoughtful understanding of the topic at hand as well as of the goals and purposes of carrying out particular mathematical processes.

Teachers must also consider the students in deciding on the appropriateness of a given task. They must consider what they know about their particular students as well as what they know more generally about students from psychological, cultural, sociological, and political perspectives. For example, teachers should consider gender issues in selecting tasks, deliberating about ways in which the tasks may be an advantage either to boys or to girls—and a disadvantage to the others—in some systematic way.

In thinking about their particular students, teachers must weigh several factors. One centers on what their students already know and can do, what they need to work on, and how much they seem ready to stretch intellectually. Well-chosen tasks afford teachers opportunities to learn about their students' understandings even as the tasks also press the students forward. Another factor is their students' interests, dispositions, and experiences. Teachers should aim for tasks that are likely to engage their students' interests. Sometimes this means choosing familiar application contexts: for example, having students explore issues related to the finances of a school store or something in the students' community. Not always, however, should concern for "interest" limit the teacher to tasks that relate to the familiar everyday worlds of the students; theoretical or fanciful tasks that challenge students intellectually are also interesting: number theory problems, for instance. When teachers work with groups of students for whom the notion of "argument" is uncomfortable or at variance with community norms of interaction, teachers must consider carefully the ways in which they help students to engage in mathematical discourse. Defensible reasoning about students must be based on the assumption that all students can learn and do mathematics, that each one is worthy of being challenged intellectually. Sensitivity to the diversity of students' backgrounds and experiences is crucial in selecting worthwhile tasks.

Knowledge about ways in which students learn mathematics is a third basis for appraising tasks. The mode of activity, the kind of thinking required, and the way in which students are led to explore the particular content all contribute to the kind of learning opportunity afforded by the task. Knowing that students need opportunities to model concepts concretely and pictorially, for example, might lead a teacher to select a task that involves such representations. An awareness of common student confusions or misconceptions around a certain mathematical topic would help a teacher to select tasks that engage students in exploring critical ideas that often underlie those confusions. Understanding that writing about one's ideas helps to clarify and develop one's understandings would make a task that requires students to write explanations look attractive. Teachers' understandings about how students learn mathematics should be informed by research as well as their own experience. Just as teachers can learn more about students' understandings from the tasks they provide students, so, too, can they gain insights into how students learn mathematics. To capitalize on the opportunity, teachers should deliberately select tasks that provide them with windows on students' thinking.

The teacher analyzes the content and how to approach it, and she considers how it connects with other mathematical ideas.

1.1 Mrs. Jackson is thinking about how to help her students learn about perimeter and area. She realizes that learning about perimeter and area entails developing concepts, procedures, and skills. Students need to understand that the perimeter is the distance around a region and the area is the amount of space inside the region and that length and area are two fundamentally different kinds of measure. They need to realize that perimeter and area are not directly related—that, for instance, two figures can have the same perimeter but different areas. Students also need to be able to figure out the perimeter and the area of a given region. At the same time, they should relate these to other measures with which they are familiar, such as measures of volume or weight.

Mrs. Jackson examines two tasks designed to help upper elementary-grade students learn about perimeter and area. She wants to compare what each has to offer.

TASK 1:

Find the area and perimeter of each rectangle:

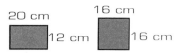

Task 1 requires little more than remembering what "perimeter" and "area" refer to and the formulas for calculating each. Nothing about this task requires students to ponder the relationship between perimeter and area. This task is not likely to engage students intellectually; it does not entail reasoning or problem solving.

TASK 2:

Suppose you had 64 meters of fence with which you were going to build a pen for your large dog, Bones. What are some different pens you can make if you use all the fencing? What is the pen with the least play space? What is the biggest pen you can make—the one that allows Bones the most play space? Which would be best for running?

This task can engage students intellectually because it challenges them to search for something. Although accessible to even young students, the problem is not immediately solvable. Neither is it clear how best to approach it. A question that students confront as they work on the problem is how to determine that they have indeed found the largest or the smallest play space. Being able to justify an answer and to show that a problem is solved are critical components of mathematical reasoning and problem solving. The problem yields to a variety of tools—drawings on graph paper, constructions with rulers or compasses, tables, calculators—and lets students develop their understandings of the concept of area and its relationship to perimeter. They can investigate the patterns that emerge in the dimensions and the relationship between those dimensions and the area. This problem may also prompt the question of what "largest" or "smallest," "most" or "least" mean, setting the stage for making connections in other measurement contexts.

1.2 Ms. Pierce is a first-year teacher in a large middle school. She uses a mathematics textbook, published about ten years ago, that her department requires her to follow closely. In the middle of a unit on fractions with her seventh graders, Ms. Pierce is examining her textbook's treatment of division with fractions. She is trying to decide what its strengths and weaknesses are and whether and how she should use it to help her students understand division with fractions.

She notices that the textbook's emphasis is on the mechanics of carrying out the procedure ("dividing by a number is the same as multiplying by its reciprocal"). The text tells students that they "can use reciprocals to help" them divide by fractions and gives them a few examples of the procedure.

The picture at the top of one of the pages shows some beads of a necklace lined up next to a ruler—an attempt to represent, for example, that there are twenty-four ¾-inch beads and forty-eight ⅜-inch beads in an eighteen-inch necklace. Ms. Pierce sees that this does represent what it means to divide by ¾ or by ⅜—that the question is, "How many three-fourths or three-eighths are there in eighteen?" Still, when she considers what would help her students understand this, she does not think that this representation is adequate. She also suspects that students may not take this section seriously, for they tend to believe that mathematics means memorizing rules rather than understanding why the rules work.

Ms. Pierce is concerned that these pages are likely to reinforce that impression. She doesn't see anything in the task that would emphasize the value of understanding why, nor that would promote mathematical discourse.

Thinking about her students, Ms. Pierce judges that these two pages require computational skills that most of her students do have (i.e., being able to produce the reciprocal of a number, being able to multiply fractions) but that the exercises on the pages would not be interesting to them. Nothing here would engage their thinking.

Looking at the pictures of the necklaces gives Ms. Pierce an idea. She decides that she can use this idea, so she copies the drawing only. She will include at least one picture with beads of some whole number length—2-inch beads, for example. She will ask students to examine the pictures and try to write some kind of number sentence that represents what they see. For example, this 7-inch bracelet has 14 half-inch beads:

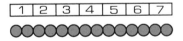

This could be represented as $7 ÷ ½$ or $7 × 2$. She will try to help them to think about the reciprocal relationship between multiplication and division and the meaning of dividing something by a fraction or by a whole number. Then, she thinks, she could use some of the exercises on the second page but, instead of just having the students compute the answers, she will ask them, in pairs, to write stories for each of about five exercises.

She decides she will also provide a couple of other examples that involve whole number divisors: $28 ÷ 8$ and $80 ÷ 16$, for example.

Many beginning and experienced teachers are in the same position as this teacher: having to follow a textbook quite closely. Appraising and deciding how to use textbook material is critical.

The teacher wants her students to understand what it means to divide by a fraction, not just learn the mechanics of the procedure.

The teacher senses that the idea of "using the reciprocal" is introduced almost as a trick, lacking any real rationale or connection to the pictures of necklaces. Furthermore, division with fractions seems to be presented as a new topic, unconnected to anything the students might already know, such as division of whole numbers.

The practice exercises involve dividing one fraction by another, and the "problems" at the end do not involve reasoning or problem solving.

The teacher considers what she knows about her students—what they know and what is likely to interest them.

The model used is a linear one rather than the pie or pizza diagrams most often used to represent fractions. The teacher sees the need for students to develop varied representations. Also, different representations make sense to different students. The teacher wants the task to help students make connections—in this example, between multiplication and division and between division of whole numbers and division of fractions.

Writing stories to go with the division sentences may help students to focus on the meaning of the procedure.

The teacher keeps her eye on the bigger curricular picture as she selects and adapts tasks. Juxtaposing whole number and fraction division will help her students review division and make connections.

Ms. Pierce feels encouraged from her experience with planning this lesson and thinks that revising other textbook lessons will be feasible. Despite the fact that she is supposed to be following the text closely, Ms. Pierce now thinks that she will be able to adapt the text in ways that will significantly improve what she can do with her students this year.

1.3 After recently completing a unit on multiplication and division, a fourth-grade class has just begun to learn about factors and multiples. Their teacher is using the calculator as a tool for this topic. This approach is new for her. The school has just purchased for the first time a set of calculators, which all the classrooms share. She and many of her colleagues attended a workshop recently on different uses of calculators.

The teacher uses this exploratory task to spur students' mathematical thinking. She knows that the initial task is likely to generate further, more focused tasks based on the students' conjectures. The calculators help the students in looking for patterns.

Using the automatic constant feature of their calculators (that is, that pressing 5 + = = =...yields 5, 10, 15, 20,...on the display), the fourth graders have generated lists of the multiples of different numbers. They have also used the calculator to explore the factors of different numbers. To encourage the students to deepen their understanding of numbers, the teacher has urged them to look for patterns and to make conjectures. She asked them, "Do you see any patterns in the lists you are making? Can you make any guesses about any of those patterns?"

All year, this teacher has encouraged her students to take intellectual risks by asking questions.

Two students have raised a question that has attracted the interest of the whole class:

Are there more multiples of 3 or more multiples of 8?

Judging that this question is a fruitful one, the teacher picks up on the students' idea and uses it to further the direction of the class's exploration, even bringing up questions about infinity.

The teacher encourages them to pursue the question, for she sees that this question can engage them in the concept of multiples as well as provide a fruitful context for making mathematical arguments. She realizes that the question holds rich mathematical potential and even brings up questions about infinity. "What do the rest of you think?" she asks. "How could you investigate this question? Go ahead and work on this a bit on your own or with a partner and then let's discuss what you come up with."

The question promotes mathematical reasoning, eliciting at least three competing and, to fourth graders, compelling mathematical arguments. Students are actively engaged in trying to persuade other members of the class of the validity of their argument.

The children pursue the question excitedly. The calculators are useful once more as they generate lists of the multiples of 3 and the multiples of 8. Groups are forming around particular arguments. One group of children argues that there are more multiples of 3 because in the interval between 0 and 20 there are more multiples of 3 than multiples of 8. Another group is convinced that the multiples of 3 are "just as many as the multiples of 8 because they go on forever." A few children, thinking there should be more multiples of 8 because 8 is greater than 3, form a new conjecture about numbers—that the larger the number, the more *factors* it has.

The teacher is pleased with the ways in which opportunities for mathematical reasoning are growing out of the initial exploration. She likes the way in which they are making connections between multiples and factors. She also notes that students already seem quite fluent using the terms *multiple* and *factor*.

The task has stimulated students to formulate a new problem. The idea that lessons can raise questions for students to pursue is part of an emphasis on mathematical inquiry.

The teacher provides a context for dealing with students' conjectures. She is also able to formulate tasks out of the students' ideas and questions when it seems fruitful.

Although it is nearing the end of class, the teacher invites them to present to the rest of the class their conjecture that the larger the number, the more factors it has. She suggests that the students record it in their notebooks and discuss it in class tomorrow. Pausing for a moment before she sends them out to recess, she decides to provoke their thinking a little and remarks, "That's an interesting conjecture. Let's just think about it for a sec. How many factors does, say, 3 have?"

"Two," call out several students.

"What are they?" she probes. "Yes, Deng?"

"1 and 3," replies Deng quickly.

"Let's try another one," continues the teacher. "What about 20?"

After a moment, several hands shoot up. She pauses to allow students to think and asks, "Natasha?"

"Six—1 and 20, 2 and 10, 4 and 5," answers Natasha with confidence.

The teacher throws out a couple more numbers—9 and 15. She is conscious of trying to use only numbers that fit the conjecture. With satisfaction, she notes that most of the students are quickly able to produce all the factors for the numbers she gives them. Some used paper and pencil, some used calculators, and some did a combination of both. As she looks up at the clock, one child asks, "But what about 17? It doesn't seem to work."

"That's *one* of the things that you could examine for tomorrow. I want all of you to see if you can find out if this conjecture always holds."

"I don't think it'll work for odd numbers," says one child.

"Check into it," smiles the teacher. "We'll discuss it tomorrow."

The teacher provides practice in multiplication facts at the same time that she engages the students in considering their peers' conjecture.

The teacher does not want to give them a key to challenging the conjecture, but she does want to get them into investigating it.

She tries to spur them on to pursuing this idea on their own.

The teacher deliberately leaves the question unanswered. She wants to encourage them to persevere and not expect her to give the answers.

Summary: Tasks

The teacher is responsible for shaping and directing students' activities so that they have opportunities to engage meaningfully in mathematics. Textbooks can be useful resources for teachers, but teachers must also be free to adapt or depart from texts if students' ideas and conjectures are to help shape teachers' navigation of the content. The tasks in which students engage must encourage them to reason about mathematical ideas, to make connections, and to formulate, grapple with, and solve problems. Students also need skills. Good tasks nest skill development in the context of problem solving. In practice, students' actual opportunities for learning depend on the kind of *discourse* that the teacher orchestrates, an issue we examine in the next section.

DISCOURSE

The discourse of a classroom—the ways of representing, thinking, talking, agreeing and disagreeing—is central to what students learn about mathematics as a domain of human inquiry with characteristic ways of knowing. Discourse is both the way ideas are exchanged and what the ideas entail: Who talks? About what? In what ways? What do people write, what do they record and why? What questions are important? How do ideas change? Whose ideas and ways of thinking are valued? Who determines when to end a discussion? The discourse is shaped by the tasks in which students engage and the nature of the learning environment; it also influences them.

Discourse entails fundamental issues about knowledge: What makes something true or reasonable in mathematics? How can we figure out whether or not something makes sense? That something is true because the teacher or the book says so is the basis for much traditional classroom discourse. Another view, the one put forth here, centers on mathematical reasoning and evidence as the basis for the discourse. In order for students to develop the ability to formulate problems, to explore, conjecture, and reason logically, to evaluate whether something makes sense, classroom discourse must be founded on mathematical evidence.

Students must talk, with one another as well as in response to the teacher. When the teacher talks most, the flow of ideas and knowledge is primarily from teacher to student. When students make public conjectures and reason with others about mathematics, ideas and knowledge are developed collaboratively, revealing mathematics as constructed by human beings within an intellectual community. Writing is another important component of the discourse. Students learn to use, in a meaningful context, the tools of mathematical discourse—special terms, diagrams, graphs, sketches, analogies, and physical models, as well as symbols.

The teacher's role is to initiate and orchestrate this kind of discourse and to use it skillfully to foster student learning. In order to facilitate learning by all students, teachers must also be perceptive and skillful in analyzing the culture of the classroom, looking out for patterns of inequality, dominance, and low expectations that are primary causes of nonparticipation by many students. Engaging every student in the discourse of the class requires considerable skill as well as an appreciation of, and respect for, students' diversity.

STANDARD 2:
THE TEACHER'S ROLE IN DISCOURSE

The teacher of mathematics should orchestrate discourse by—

♦ *posing questions and tasks that elicit, engage, and challenge each student's thinking;*

♦ *listening carefully to students' ideas;*

♦ *asking students to clarify and justify their ideas orally and in writing;*

♦ *deciding what to pursue in depth from among the ideas that students bring up during a discussion;*

♦ *deciding when and how to attach mathematical notation and language to students' ideas;*

♦ *deciding when to provide information, when to clarify an issue, when to model, when to lead, and when to let a student struggle with a difficulty;*

♦ *monitoring students' participation in discussions and deciding when and how to encourage each student to participate.*

Elaboration

Like a piece of music, the classroom discourse has themes that pull together to create a whole that has meaning. The teacher has a central role in orchestrating the oral and written discourse in ways that contribute to students' understanding of mathematics.

The kind of mathematical discourse described above does not occur spontaneously in most classrooms. It requires an environment in which everyone's thinking is respected and in which reasoning and arguing about mathematical meanings is the norm. Students, used to the teacher doing most of the talking while they remain passive, need guidance and encouragement in order to participate actively in the discourse of a collaborative community. Some students, particularly those who have been successful in more traditional mathematics classrooms, may be resistant to talking, writing, and reasoning together about mathematics.

One aspect of the teacher's role is to provoke students' reasoning about mathematics. Teachers must do this through the tasks they provide and the questions they ask. For example, teachers should regularly follow students' statements with, "Why?" or by asking them to explain. Doing this consistently, irrespective of the correctness of students' statements, is an important part of establishing a discourse centered on mathematical reasoning. Cultivating a tone of interest when asking a student to explain or elaborate on an idea helps to establish norms of civility and respect rather than criticism and doubt. Teachers also stimulate discourse by asking students to write explanations for their solutions and provide justifications for their ideas.

Emphasizing tasks that focus on thinking and reasoning serves to provide the teacher with ongoing assessment information. Well-posed questions can simultaneously elicit and extend students' thinking. The teacher's skill at formulating questions to orchestrate the oral and written discourse in the direction of mathematical reasoning is crucial.

A second feature of the teacher's role is to be active in a different way from that in traditional classroom discourse. Instead of doing virtually all the talking, modeling, and explaining themselves, teachers must encourage and expect students to do so. Teachers must do more listening, students more reasoning. For the discourse to promote students' learning, teachers must orchestrate it carefully. Because many more ideas will come up than are fruitful to pursue at the moment, teachers must filter and direct the students' explorations by picking up on some points and by leaving others behind. Doing this prevents student activity and talk from becoming too diffuse and unfocused. Knowledge of mathematics, of the curriculum, and of students should guide the teacher's decisions about the path of the discourse. Other key decisions concern the teacher's role in contributing to the discourse. Beyond asking clarifying or provocative questions, teachers should also, at times, provide information and lead students. Decisions about when to let students struggle to make sense of an idea or a problem without direct teacher input, when to ask leading questions, and when to tell students something directly are crucial to orchestrating productive mathematical discourse in the classroom. Such decisions depend on teachers' understandings of mathematics and of their students—on judgments about the things that students can figure out on their own or collectively and those for which they will need input.

A third aspect of the teacher's role in orchestrating classroom discourse is to monitor and organize students' participation. Who is volunteering comments and who is not? How are students responding to one another? What are different students able to record or represent on paper about their thinking? What are they able to put into words, in what kinds of contexts? Teachers must be committed to engaging every student in contributing to the thinking of the class. Teachers must judge when students should work and talk in small groups and when the whole group is the most useful context. They must make sensitive decisions about how turns to speak are shared in the large group—for example, whom to call on when and whether to call on particular students who do not volunteer. Substantively, if the discourse is to focus on making sense of mathematics, on learning to reason mathematically, teachers must refrain from calling only on students who seem to have right answers or valid ideas to allow a broader spectrum of thinking to be explored in the discourse. By modeling respect for students' thinking and conveying the assumption that students make sense, teachers can encourage students to participate within a norm that expects group members to justify their ideas. Teachers must think broadly about a variety of ways for students to contribute to the class's thinking—using means that are written or pictorial, concrete or representational, as well as oral.

Vignettes

2.1 Ms. Nakamura has done a lot more number work with her kindergarten class this year, and she is pleased with how this is going. Now, near the end of the year, they have been investigating patterns in the number of various body parts in the classroom—how many noses or eyes, for example, there are among the children in the class.

Earlier this week, each child made a nose out of clay. Ms. Nakamura opens the discussion by revisiting this project. She asks: And how many noses did we make?

This task creates an opportunity for counting and reasoning about numbers. Some body parts are double the number of people in the room, whereas others are identical to the number of people. Some body parts (such as fingers, for example) would be still more.

Becky (points to her nostrils): Two of these.

Teacher: But how many actual *noses*?

Anne: 29.

Teacher: Why? *Why* were there 29 noses?

Adam: Because every kid in the class made one clay nose and that is the same number as kids in the class.

Teacher (pointing to her nostrils): Now Becky just said—remember what these are called?

Children: Nostrils!

Teacher: So were there 29 nostrils?

Pat: No, there were more.

Gwen: 58! We had 58 nostrils!

Teacher: Why 58?

Gwen: I counted.

Felice: If we had 30 kids, we would be 60. So it is 59 'cause it should be one less.

Teacher: Can you explain that again?

Felice: It's 59 because we don't have 30 kids, we have 29, so it is one less than 60.

Teacher: What does anyone else think?

Adam: I think it is 58. Each kid has *2* nostrils. So if 60 would be for 30 kids, then it has to be two less: 58.

Lawrence: But Felice says 30 kids makes 60....

Felice: No! That makes sense. 58.

The teacher looks around at the children, some of whom are beginning to wriggle. She waits, and then asks: What do the rest of you think?

Two girls chorus: It's 58.

Several others join: 58.

Teacher: So it's 58 because 30 kids would have 60 nostrils and we have to take away 2 for one less kid. Gwen said she counted. Let's count and see.

Ms. Nakamura leads the class through counting nostrils: 1, 2. 3, 4. 5, 6....

She moves on: What else do you think we have on our bodies that would be more than 29?

Graham: More than 29 fingers.

Teacher: More than 29 fingers? Why do you think so?

Graham: Because each kid, we have 10 fingers.

Ricky: More than 29 shoes.

Teacher: More than 29 shoes. And what are those shoes covering?

Ricky: Your feet.

Sarah: Ears.

Beth: More than 29 legs.

Ricky: And that goes with the feet idea.

Teacher: And why do you think that goes with the feet idea?

The teacher consistently asks students to explain and to justify their answers. "Why?" is a standard question, asked about apparently correct as well as about apparently wrong answers.

The teacher probes Felice's answer even though this goes beyond what many of the children are trying to do at this point.

The teacher solicits other students' reactions instead of showing them the right answer. Her tone of voice and her questions show the students that she values their thinking.

She allows time for children to think and remains neutral about the correctness of what is being said. She would like them to monitor whether mathematical ideas make sense by reasoning about them.

Here the teacher summarizes what different children have contributed to investigating Gwen's answer. Because Gwen's reasoning is complicated, the teacher then leads them through another means of verifying the result.

The teacher's question challenges students to think. It is open-ended; more than one right answer exists. Consider the difference between her question here and asking, "Do we have more than 29 fingers?"

Here the teacher lets the pace pick up by allowing these suggestions one right after the other without probing them for their explanations. Still, it might have helped children to hear one another's reasoning. Decisions should take into consideration when to move quickly and when to make sure an idea is thoroughly justified, when to pursue additional issues and when to remain focused on a particular purpose.

Ricky: 'Cause the feet are attached to the shoes.

Teacher: Your shoes. But you said that Beth's idea went with your idea. Why does her leg idea go with your idea?

Ricky: 'Cause you put them on over your legs. Because your feet are attached to your legs.

Teacher: Oh, so your feet are attached to your legs.

Willie: Legs are attached to knees.

Teacher: Legs are attached to knees, so your idea of knees, Beth's idea of legs, and Ricky's idea of feet—they all kind of go together, don't they? They're all attached.

Paul: The feet are attached to the legs, the legs are attached to the knees, the knees are attached to the thighs, and the thighs are attached to the chest.

The teacher chooses to overlook the comment that the thighs are connected to the chest because she is focusing on children's one-to-one reasoning.

Teacher: And how many chests are there altogether?

Children: 29.

Liza: Because we have one chest.

Recording their ideas and providing a representation of their reasoning (justification) help develop students' capacity with the written aspects of discourse.

Ms. Nakamura tells the children that they are to work on a picture now: Choose some body part and draw a picture of how many of those we have in our class and how you *know* that.

The teacher chooses to comment on the children's thinking instead of their behavior.

She directs them back to their tables where she has laid out paper and cans of crayons: You did some good thinking today!

The teacher elicits students' ideas in beginning a discussion. She is able to gather information about what students know and assume that it can guide her subsequent questions.

2.2 Mrs. Logan is beginning a geometry unit with her students. She opens class by announcing: We're going to be studying about quadrilaterals. What do you know about quadrilaterals?

Several students chorus: Four sides, four-sided figure.

Mrs. Logan draws

Reacting directly to what students say, the teacher provokes their thinking by building on the descriptions they have given.

and asks: Is this one?

Students: No, it has to connect.

Mrs. Logan: Is this one?

and asks: Is this one?

Several students: No, it can't intersect like that.

Each time she uses what they have said to construct her next response. This reflects the tight connection between students' thinking and her moves in orchestrating the discourse.

Mrs. Logan continues drawing and asks: So is this one?

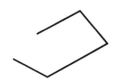

Student: It has to close.

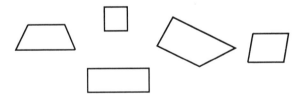

Mrs. Logan: Okay, then is this one?

Students: Yes!

Mrs. Logan pauses and then looks directly at the students: I drew four examples. You said three of these didn't work. Can you explain what makes the difference?

The teacher asks a question designed to stimulate students to formulate the pieces of what they have been saying about quadrilaterals.

Several students volunteer pieces of a definition of "quadrilateral." Mrs. Logan lists their ideas—in their terms—on the board:

QUADRILATERALS

4 points
4 segments
no more points intersect
closed curve

Summarizing, Mrs. Logan tells them: You really have all the pieces. The definition in our text is "the union of segments joining four points such that the segments intersect only at the endpoints."

The teacher connects students' ideas with the mathematical definition and decides to provide some additional information.

Now, which of our figures fit with this definition?

After a short discussion of their figures, Mrs. Logan continues: And what are some special kinds of four-sided figures?

Again, she asks a question designed to elicit the students' knowledge as a means of developing the class's consideration of the topic.

Students call out a variety of names: square, rectangle, kite, rhombus, parallelogram.

Mrs. Logan: Another one you should have heard of before is trapezoid.

The teacher adds to their contributions as needed.

She draws several figures on the board: Can you classify these and talk about them?

Students begin murmuring, in pairs and threes.

She provides a context for them to discuss quadrilaterals among themselves.

Mrs. Logan: Are there other special quadrilaterals that are not up here?

She adds a question to provoke their conversations further.

One usually reticent student asks: But isn't a square also a rectangle? I don't quite see how to classify these.

Mrs. Logan replies: Why don't you put that to the rest of the class? See what they think?

The student repeats his question to his classmates. Mrs. Logan adds: See if you can find a way to classify these shapes. Which shapes have which labels?

Instead of responding directly to his question, the teacher decides it is worth everyone's consideration and redirects it to the group. She also sees it as a way of stimulating this particular student's participation in the classroom discourse.

The students resume their work. Some begin making diagrams to represent the interrelationship among the types of quadrilaterals, others are making tables. Mrs. Logan walks around, asking students questions to get them to clarify what they are thinking. She asks one boy to explain his rather complicated chart: And why is the rhombus there, with the parallelogram?

After class, Mrs. Logan contemplates the lesson. In general, she feels it was a good start. Perhaps tomorrow—in order to help the students begin to construct some of the categories—she will engage them in some kind of problem or activity in which they will have to sort quadrilaterals. She muses a bit about how to frame it in a way that will promote discourse.

2.3 Mr. Luu has been working on probability for a few days with his class of sixth graders. Because his textbook is old, there is little about probability in the book. He has been drawing from a variety of sources as well as making up things himself, based on what he hears in the students' comments. He began by asking students to decide whether a coin-tossing game he presented was fair or not. He found out that although most of the students did consider the possible outcomes, they did not analyze the ways those outcomes could be obtained. For example, they thought that when you toss two coins, it is equally likely to get two heads, two tails, or heads-and-tails. He also learned that many of his students were inclined to decide if a game was fair by playing it and seeing if the players tied: If someone won, then the game might be biased in their favor, they thought.

He decides to present them with two dice-tossing games—the sum game and the product game:

SUM GAME

Two players:
Choose one player to be "even" and the other to be "odd."
Throw two dice.
Add the numbers on the two faces.
If the sum is even, the even player gets 1 point.
If the sum is odd, the odd player gets 1 point.

PRODUCT GAME

Two players:
Choose one to be the "even" player and the other to be "odd."
Throw two dice.
Multiply the numbers on the two faces.
If the product is even, the even player gets 1 point.
If the product is odd, the odd player gets 1 point.

The teacher poses questions and problems that both elicit and challenge students' thinking.

After explaining how each game is played, Mr. Luu challenges the students to figure out if the games are fair or not. He begins by holding a discussion about what it means for something to be "fair." Then he presents the rules for each game, telling the students simply that they are to report back on whether or not either of the games is fair or not and to include an explanation for their judgment.

The teacher makes a decision about how much to focus a problem, how much to direct the students. Here, he decides that a common understanding of what makes something "fair" is crucial.

The students pair off and work on the problem. Some play each of the games first, recording their results, as a means of investigating the question. Others try to analyze the games based on the possible outcomes. Mr. Luu walks around and listens to what the students are saying and poses questions:

"What did you say were all the possible totals you could get? How did you know?"

"Why did you decide you needed to throw the dice exactly 36 times?"

The teacher tries to provoke students' thinking. For example, he knows that the students who planned to throw the dice exactly 36 times may be assuming that the experimental results should be the same as their predicted outcomes.

After they have played the game or worked on their analyses for a while, Mr. Luu directs the students to stop, to open their notebooks, and to write in their notebooks what they think about the fairness of the two games.

The teacher appreciates the importance of writing about mathematics, and he provides regular occasions for it.

Next, Mr. Luu opens a whole-class discussion about the games. On the basis of what he saw when he was observing, he calls on Kevin and Rania. Rania beams. She explains that they figured out that the sum game is an unfair game "and we didn't even have to play it at all to be sure."

Kevin provides their proof: "There are six even sums possible—2, 4, 6, 8, 10, and 12—but only five odd ones—3, 5, 7, 9, and 11. So the game is unfair to the person who gets points for the odd sums."

Mr. Luu has called on these two students because on the one hand, they are comfortable with the idea that probabilities can be analyzed, that the game need not actually be played. But on the other hand, these students have made an erroneous conclusion. He thinks that this combination makes their solution a good lead-off for the whole-group discussion.

The teacher expects the students to evaluate Kevin and Rania's argument and to decide together whether or not it makes sense.

The teacher makes careful decisions about when and how to encourage each student to participate. This time he is rewarded; sometimes when he calls on someone in this way, he gets stony silence.

Instead of explaining what Marcus has said, the teacher expects Marcus to provide his own clarification and justification.

Mr. Luu decides to press this issue, for he knows that understanding the concept of "outcome" is central to understanding probability. He thinks they can resolve this themselves, so he nudges the discussion along.

"What do the rest of you think?" asks Mr. Luu, gazing over the group. Several shake their heads. A few others nod.

"Marcus?" he invites. Marcus's hand was not up, but his face looks up at Mr. Luu. "It don't make sense to me, Mr. Luu. I think that there's more *ways* to get some of them numbers, like 3—there's two ways to get a 3. But there is only one way to get a 2."

"Huh?" Several children are openly puzzled by this statement.

"Marcus, can you explain what you mean by saying that 3 can be made two ways?" asks Mr. Luu.

"Well, you could get a 1 on one die and a 2 on the other, *or* you could get a 2 on the first die and a 1 on the other. That's two different ways," he explains quietly.

"But how are those different? One plus two equals the same thing as two plus one!" objects a small girl.

"What do you think, Than?" probes Mr. Luu.

Than remains silent. Mr. Luu waits a long time. Finally Than says, "But they are two different dices, so it is *not* same."

"Hmmm," remarks Mr. Luu. "Where are other people on this?"

After three or four more comments on both sides of the issue, time is almost up. Mr. Luu assigns the students, for homework, to repeat the coin-tossing game they had investigated last week, to record their results, and to decide if it is fair when three people play it:

COIN-TOSSING GAME

Three players:
One player is "two heads," one player is "two tails,"
and one player is "mixed."

Toss two coins.

If the result is two heads, the "heads" player gets 1 point.
If the result is two tails, the "tails" player gets 1 point.
If the result is one head and one tail, the "mixed" player gets 1 point.

The teacher extends the discouse by assigning a writing task coupled with more data collection.

Mr. Luu judges that, mathematically, this is an appropriate question for his students to struggle with. It is closely tied with the dice problem but may make the key concept of "outcome" more accessible.

Mr. Luu thinks that this game may help them with their thinking about the dice games. He asks them to play the game, to record their results, and to decide if it is fair when three people play it. They are to write about their experiments and explain their conclusion. Mr. Luu suspects that now, if they find out that the "mixed" result person gets about twice as many points as either of the others, they will be able to figure out what is going on and eventually agree with Marcus and Than.

2.4 Toward the end of a unit on quadratic equations, Mr. Santos has decided to assess his algebra students' use of problem-solving processes and their ability to make mathematical connections, both among ideas in the unit and between these ideas and concepts covered earlier. To do this, he chooses the following problem from the 1988 NCTM Yearbook, *The Ideas of Algebra, K–12* (p. 19):

Find all the values of x for which

$$(x^2 - 5x + 5)^{x^2 - 9x + 20} = 1.$$

He decides to ask students to work on the problem in pairs while he circulates among as many of the pairs as he can, monitoring their progress. He uses a checklist with students' names on it as an easy means of recording observations about students' thinking, approaches, and patterns of working.

The first pair of students he visits, Alan and Bettina, groan, "This is really going to be gross!" "Look, it's got two different quadratics in the same equation!" "Yeah, it's not fair. He never gave us such a complicated one before!" "Oh, well," Bettina says, "we might as well get started. Let's factor $x^2 - 9x + 20$ and see what we get." When they find that $(x - 4)$ and $(x - 5)$ are the factors, Alan says, "Well, I guess that's it, $x = 4$ and $x = 5$ must be the answers."

Bettina does not seem certain about Alan's assertion. "What about this other quadratic? Don't we have to check that it works there, too?" she asks. "Oh, yeah," agrees Alan, "you check 5 and I'll check 4." So, they substitute 4 and 5 into $x^2 - 5x + 5$ and find that they get 1 for $x = 4$ and they get 5 for $x = 5$. Alan says, "I get 1 like I'm supposed to," but Bettina says, "I don't get 1, I get 5." This result puzzles them.

As they look at the problem together, Alan says, "We need to use both quadratics together," and Bettina chimes in, "Yeah, it's this to that power." Evaluating the entire expression, they find that for $x = 4$ they get 1^0 and for $x = 5$ they get 5^0. They comment that "it's one either way; anything to the 0 power is 1."

Alan leans back, seeming confident that they have solved the problem. Bettina, still engaged with the problem, says, "Hey, look, if this is 1, then the exponent could be anything. Can we use that?"

Taking up Bettina's question, Alan points at the base, $x^2 - 5x + 5$, and says, "Okay, you mean we should see if any other values of x could make this part equal to 1?"

Out of the corner of his eye, Mr. Santos notices a pair of students clowning around by the window. He hears them laughing and sees them pushing one another playfully. He approaches them and asks, "What's up?"

"No way we can do this problem, Mr. S," says Diarra.

"And, besides, who CARES?" adds Tommy.

Mr. Santos guesses that part of their frustration is that the problem looks too complex. He invites them to try the problem by putting in some numbers.

"How about 1?" suggests Diarra, giggling.

"Yeah," agrees Tommy.

When they try 1, they are surprised to see that it works out.

He notices that it is headed "*Can your Algebra Class Solve This?*" and recognizes that it is likely to be a tough one for them, but he expects that as a nonroutine problem it will serve as an alternative means of assessment. He hopes to gain insight into the students' learning and development of mathematical disposition.

The teacher's skill in orchestrating discourse is enhanced by close knowledge of students.

The teacher does not take these complaints too seriously, noting that these students exhibit an improving mathematical disposition by getting down to work on the problem.

The teacher suspects that these students may be getting a right answer for a wrong reason. But since they seem about to verify their solution, he decides to observe and listen to them a little longer to see what happens.

The teacher notices with pleasure that Bettina seems persistent, reflective, and on the lookout for additional solutions. He notes that Alan expects to reach the answer quickly and is satisfied with a single answer. He thinks that it would be worthwhile for Alan to engage in more tasks of this sort and makes a mental note to think further about how to develop students' persistence and reflectiveness. Alan and Bettina have a new idea to work on, and Mr. Santos moves on to observe another group.

The teacher monitors students' engagement in the mathematics. He communicates that he expects them to be involved.

The teacher decides to give them a hint to head them toward solving the problem. He expects that this will help to get them involved.

◆   ◆   ◆   ◆   ◆   ◆   ◆   ◆

"Hey, this is easy, man!" exclaims Tommy. At this, other students crowd around.

The teacher asks a question to press the students onward with solving the problem.

"Are there other solutions?" asks Mr. Santos, relieved that the students are becoming engaged.

The teacher's efforts to engage the students are paying off.

"I'll try 2," volunteers one. Others are trying other numbers. As he walks away, Mr. Santos hears another burst of excitement as a pair of students discovers that 2 works also He also hears a groan from a student who has tried 0.

The teacher is pleased with these students' ability to express their mathematical thinking and problem-solving strategies, and he looks forward to reading their entries in the mathematics journals the students have been keeping.

Looking around the classroom, Mr. Santos notices a pair of students, Geraldo and Linnea, using a graphing calculator. When he goes over to them, they tell him that they have graphed the functions $y = 1$ and

$$y = (x^2 - 5x + 5)^{x^2 - 9x + 20}$$

and are now zooming in to look for the points of intersection. When he asks them to explain what they have been doing, they say that they decided from the beginning to use a graphing calculator. They describe moving from graphing the two quadratics separately to using the exponentiation key and graphing the whole function at once. Linnea says, "We finally realized that what we had here was a polynomial to a polynomial power."

The teacher decides that this problem will provide an excellent vehicle for discourse on the integration of technology and algebra to solve problems.

Mr. Santos asks them about the section of discontinuity on the graph and if their "picture" represents a complete graph. He suddenly realizes that this pair of students has provided him with additional insight into this problem and makes a mental note to change his lesson plans for later in the week. He will bring in the demonstration computer so that the whole class can participate in further discussion on using technology to solve this problem.

Mr. Santos looks around the class for another group to visit and notices another pair, Peter and Ona, lounging with nothing to do. "How are you two doing?" he inquires pleasantly.

"Great!" Peter replies, "We got the answer; it's 4 and 5." They show Mr. Santos how they did it. They have used an approach similar to the one used by Alan and Bettina.

The teacher wants to provoke them to continue looking for solutions; he decides to ask a question that he hopes will challenge their idea that they have finished and extend their thinking about the problem.

Mr. Santos asks, "Didn't you just say that when $x = 4$ you got this polynomial [pointing to the base, $(x^2 - 5x + 5)$] to be equal to 1?" He pauses, hoping that they will notice the importance of the base having the value 1.

The teacher knows that this is a common error and makes a mental note to check the other groups to see whether they are having the same difficulty. It may be useful to allow this to surface in the whole-group discussion.

After some consideration, Peter says, "Yes, but we were worried more about the exponent being 0; but if the base is 1, the exponent wouldn't have to be 0." Ona says, "Okay, let's see if we can solve $x^2 - 5x + 5 = 1$. So they set out to factor the $x^2 - 5x + 5$, ignoring the fact that it is not set equal to 0.

Mr. Santos glances at his watch and sees that the period is almost over. He decides to end the class by reminding the students to write their journal entries for the day. They are to record the problems and successes they encountered during the period, any new insights, and anything that stood out to them about other students' arguments or solutions in class. Mr. Santos also tells the students that there are more than two solutions to the problem and that they will have another period to work on the problem on their own before class discussion of the problem takes place.

STANDARD 3:
STUDENTS' ROLE IN DISCOURSE

The teacher of mathematics should promote classroom discourse in which students—

♦ *listen to, respond to, and question the teacher and one another;*

♦ *use a variety of tools to reason, make connections, solve problems, and communicate;*

♦ *initiate problems and questions;*

♦ *make conjectures and present solutions;*

♦ *explore examples and counterexamples to investigate a conjecture;*

♦ *try to convince themselves and one another of the validity of particular representations, solutions, conjectures, and answers;*

♦ *rely on mathematical evidence and argument to determine validity.*

Elaboration

The nature of classroom discourse is a major influence on what students learn about mathematics. Students should engage in making conjectures, proposing approaches and solutions to problems, and arguing about the validity of particular claims. They should learn to verify, revise, and discard claims on the basis of mathematical evidence and use a variety of mathematical tools. Whether working in small or large groups, they should be the audience for one another's comments—that is, they should speak to one another, aiming to convince or to question their peers. Above all, the discourse should be focused on making sense of mathematical ideas, on using mathematical ideas sensibly in setting up and solving problems.

Vignettes

3.1 It is late September in a sixth-grade class. Mrs. Fondant wants to engage her students in a problem that will yield multiple solutions to help break down their image of mathematics as a domain of single right answers. One aim she has right now is to establish the norms and routines of discourse in the class. She knows that much of what she does is different from what the students have grown accustomed to in previous grades. Therefore, Mrs. Fondant takes this aspect of her task at the beginning of the year seriously. She thinks she is finally making some progress with this group, after a month of concentrating on this dimension of her work with them.

Establishing norms of discourse such as those described in this section is hard work, especially with older students who have become accustomed to a different set of standards for school thinking and talking.

She writes the following problem on the board:

The Wolverines scored 30 points in the first half of last night's basketball game. The unusual thing is that they did it without scoring a single foul shot. How did they score the 30 points?

Students are eager to share answers.

Immediately, one student yells, "That's easy! They scored ten 3-point shots!"

Another quickly interjects, "There are a lot of possibilities. It could be fifteen 2-point shots." Mrs. Fondant has one of the students explain scoring possibilities.

Students are using a variety of tools to work on the problem. In this example, solution strategies vary, but answers are the same.

Several students offer possible answers. The teacher directs them to work alone or with a partner to figure out as many answers as they can. Walking around, she notices that some are constructing tables, others are using formulas, and still others are randomly writing down combinations as they occur to them.

After a few minutes, Mrs. Fondant asks if they are ready to talk about what they have found. They seem to be, so she asks one girl for one of the combinations that she worked out.

This student already assumes that justifying her answer is part of giving it.

The girl says, "Two 3-pointers and twelve 2-pointers—2 × 3 is 6 and 12 × 2 is 24 and 6 + 24 is 30."

Several others add possible combinations.

Students initiate strategies.

Another girl asks, "How can we keep track of all these? Let's make a table."

The teacher expects students to take a large share in the pursuit of one another's suggestions.

Mrs. Fondant looks expectantly at the group. "Does someone want to make a table on the board?"

One student comes up and makes a two-column chart:

3-point shots	2-point shots

He stands at the board recording the combinations others suggest.

3-point shots	2-point shots
8	3
10	0
0	15
4	9
2	12
6	6

Students question one another's ideas.

One student questions the solution, six 3-point shots and six 2-point shots. "That makes 33 points."

A girl explains, "6 × 3 is 18 and 6 × 2 is 12. 18 + 12 is 30."

"Oh, yeah," he says.

Another student suggests redoing the table so that the combinations are in order. "Maybe we'll see a combination we missed." The teacher asks him to come up and do that.

Looking at the revised table, one student raises her hand. "Why are there no odd numbers in the 3-point box?"

Several seconds pass. Everyone seems to be pondering this. A couple of students whisper to one another.

The teacher asks what they think about this. One girl says that maybe they missed some. A few students begin searching for other combinations. After a few moments, a student says she thinks it has something to do with the fact that an odd number times an odd number equals an odd number, but she's not sure what that tells her.

In the third row, a few students are leaning together, talking quietly. One says, "I think that's important."

Mrs. Fondant asks why it would matter.

"Because for 2-point shots, the total number of points will always be even," begins one student and then pauses.

"Oh! Is it because 30 is even? And you need two even numbers to equal an even?" bursts in one of the boys who has been talking in the little group.

"What do *you* think?" asks Mrs. Fondant.

Before class ends, Mrs. Fondant asks the students, "What if they had scored *31* points—would that have changed our table?"

3.2 Ms. Chavez has rolled the math department computer into her class for the morning and has connected it to her LCD viewer. Her 28 first-year algebra students, seated at round tables in groups of threes and fours, are working on a warm-up problem. The day before they had had a test on functions. For the warm-up to today's class, Ms. Chavez has asked students to set up a table of values and graph the function $y = |x|$.

She has chosen this problem as a way to introduce some ideas for a new unit on linear, absolute value, and quadratic functions. During the warm-up, students can be heard talking quietly to one another about the problem: "Does your graph look like a V-shape?" "Did you get two intersecting lines?" Walking around the room, Ms. Chavez listens to these conversations while she takes attendance. After about five minutes, she signals that it is time to begin the whole-group discussion.

A girl volunteers and carefully draws her graph on a large wipe-off grid board at the front of the room. As she does this, most students are watching closely, glancing down at their own graphs, checking for correspondence. A few students are seen helping others who had some difficulties producing the graph.

Students support their ideas and solutions in response to others' challenges or counterarguments.

Students search for patterns and question inconsistencies that puzzle them. They are key participants in the discourse.

The teacher's question does not endorse or dismiss this idea.

Students puzzle about the mathematical clues.

The teacher expects students to look for mathematical evidence.

Students make conjectures publicly and try to convince themselves and one another of their validity.

The teacher does not directly evaluate the correctness of students' comments. Instead she interprets the comment as a conjecture and expects the students to examine its validity. She then extends the problem, trying to foster students' habits of mathematical inquiry.

Students are expected to communicate about mathematics. They seem used to paying attention to one another's ideas and to reasoning together. They also accept responsibility for helping others.

Students use a variety of tools to reason together about mathematics. They do not rely on the teacher to initiate all ideas or to certify results.

By directing them to think through comparisons, the teacher creates a context that is likely to promote students' reasoning and communication about these functions and their graphs.

The students communicate with one another about mathematics without the teacher asking them questions or directing their comments. They also use mathematical language that they have developed through the discourse.

The teacher listens to students carefully.

Students initiate conjectures publicly. They make connections between this graphing activity and transformational geometry.

Another student suggests that they enter the function into the computer and watch it produce the graph. Several other students chime in, "Yeah!" The first girl does this, and the class watches as the graph appears on the overhead screen. It matches the graph she sketched, and the class cheers, "Way to go, Elena!"

Ms. Chavez then asks the class to sketch the graphs of $y = |x| + 1$, $y = |x| + 2$, and $y = |x| - 3$ on the same set of axes and write a paragraph that compares and contrasts the results with the graph of $y = |x|$. "Feel free to work alone or with the others in your group," she tells them.

After a few minutes, two students exclaim, "All the graphs have the same shape!"

A few other students look up. Another student observes, "They're like angles with different vertex points." "Then they're really congruent angles," adds his partner.

Ms. Chavez circulates through the class, listening to the students' discussions, asking questions, and offering suggestions. She notices one group has produced only one branch of the graphs. "Why don't you choose a few negative values for x and see what happens?" Another group asks, "What would happen if we tried $|x - 3|$?" "Try it!" urges Ms. Chavez.

The students continue working, and the conversation is lowered to murmurs once again. Then the members of one group call out, "Hey, we've got something! All these graphs are just translations of $y = |x|$, just like we learned in the unit on geometry."

"That's an interesting conjecture you have," remarks Ms. Chavez. She looks expectantly at the other students. "Do the rest of you agree?" They are still, many looking hard at their graphs. One student says, slowly, "I'm not sure I get it."

A boy in the group that made the conjecture about translations explains, "Like $y = |x| + 2$ is like $y = |x|$ moved up two spaces and $y = |x| - 3$ is moved down three spaces. It's like what Louella said about them being like angles with different vertex points."

Ms. Chavez decides to provoke the class to pursue this. She asks if anyone thinks they can graph $y = |x| + 4$ without first setting up a table of values. Hands shoot up. "Ooooh!!" Scanning the class, Ms. Chavez notices Lionel, who does not volunteer often, has his hand up. He looks pleased when she invites him to give it a try.

Lionel sketches his graph on the dry-erase board. Elena again enters the equation of the graph into the computer and the class watches as the graph is produced. The computer-generated graph verifies Lionel's attempt. Again there are cheers. Lionel gives a sweeping bow and sits down.

Ms. Chavez asks the students to write in their journals, focusing on what they think they understand and what they feel unsure about from today's lesson. They lean over their notebooks, writing. A few stare into space before beginning. She gives them about ten minutes before she begins to return the tests. She will read the journals before tomorrow's class.

At the end of the of the period, she distributes the homework that she has prepared. The worksheet includes additional practice on the concept of $y = |x| \pm c$ as well as something new, to provoke the next day's discussion: $y = |x \pm c|$.

Over the next couple of weeks, students explore linear, quadratic, and absolute value functions. Nearing the end of this unit, Mrs. Chavez decides to engage students in reflecting on and assessing how far they have come.

As she assigns homework for that evening, she announces, "I'd like each of you to write two questions that you think are fair and would demonstrate that you understand the major concepts of this unit. I'll use several of your ideas to create the test. And here's a challenge for the last part of your assignment: you just drew the graph of $f(x) = x^2 - 2x$ as a part of the review. Think about everything we've done so far this semester, and see if you can remember any ideas that will help you draw the graph of $|f(x)|$."

3.3 Ms. Pizzo has been working on fractions for a little over a week with her thirty-six students, the biggest class she has ever had. She feels that she is not connecting very well with them—the group is simply too large. Many students have a conception of fractions that they picked up last year, which is that a fraction is a certain size piece of something. For example, "one-fourth" is this:

Students acknowledge their confusions openly and take one another's ideas seriously. They also expect others to respond to questions and to explain their conjectures. They then build on one another's ideas.

The teacher spurs further discussion by inserting a question that can extend the students' exploration of the conjecture.

Students present and explain solutions to the rest of the class.

The teacher does not press for closure on these ideas simply because the period ends.

These journals give the teacher insights into students' thinking. They also offer students the opportunity to reflect on their understandings and feelings.

The teacher plans homework to strengthen students' developing ideas from class and also to extend and stretch their thinking in preparation for the next class.

Both teachers and students benefit from collaborating on assessment. Teachers can gain additional information and insights about students; students gain additional opportunities to integrate and reflect on their understanding.

Students are continually pressed to seek connections.

"Three-quarters" looks, as one student said, like "a baby carriage":

The students bring their own ideas, ways of talking, and reasoning.

Ms. Pizzo is worried, though, for her students do not seem to understand fractions as numbers, nor do they see fractions as relational to some referent whole: for example, the idea of three-fourths of eighteen makes no sense to them. Three-fourths is the baby carriage shape.

Trying to think of something that will engage them and get them talking about fractions in some other ways, Ms. Pizzo decides to give them the following problem:

The teacher thinks that because they can think about "one-half of" something in a variety of ways, they may be "launched" into another way of interpreting and making sense of one-third.

$\frac{1}{2}$ of the crayons in James's box of 12 crayons are broken.

$\frac{1}{3}$ of the crayons in Fred's box of 12 crayons are broken.

Who should be sadder and why?

The students seem interested in this. They set to work, some drawing pictures, some getting out real crayons. Ms. Pizzo overhears Hilda tell Robbie, "Look, one-half is this much—

The students are using a familiar representation of fractions to model and reason about an unfamiliar situation.

and one-third is this much—

so you can just *tell* that James should be sadder."

The two children pursue a conjecture, exploring examples and looking for counterexamples. They are trying to convince themselves of what they have discovered.

Robbie, staring at Hilda's drawing, exclaims suddenly, "Look! One-third is *smaller* than one-half, even though three is more than two!"

"Hey!" answers Hilda. "Does that work with others?"

Robbie quickly draws one-fourth. The two children look at each other, excited.

The students realize that a counterexample would change what they had found.

Steven leans over. "It doesn't work. Lookit." He draws the three-fourths baby carriage shape. "Three-fourths. It *is* bigger than one-half, and four *is* more than two."

Listening to students' conversations can yield valuable information for the teacher. By attaching the label "conjecture" to their idea, she begins to help students develop a language for the mathematical thinking they are doing.

Ms. Pizzo is enjoying overhearing this interchange. "What was your idea, your conjecture?" she asks Hilda and Robbie. She hopes that she can get them to articulate what they were noticing.

By now, several other children have wandered over to Hilda's desk, having overheard this excitement.

"I guess we were saying that, with fractions, if there is a bigger number on the bottom, the fraction is smaller. It's like *opposite*," answers Robbie.

The students are in the habit of letting mathematical evidence determine the validity of an idea or an answer.

"But now with what Steven showed us, we see we were wrong," adds Hilda.

"Let's talk about this in the whole-group discussion," suggests Ms. Pizzo, pleased both that the students are finding this interesting, that they are beginning—despite the size of the group—to produce little mathematical sparks that kindle the rest of the students' interest, *and* that they are on to a key concept in fractions. She makes a note to herself to develop some kind of task that will help students investigate their ideas about the relationships between the size of denominators and the size of fractions.

The teacher does not press to elaborate or qualify the students' partial understanding of the meaning of the denominator in relation to the size of the fraction—for instance, showing them that their conjecture would work with unit fractions. Instead, she assumes that the class as a whole can work with and clarify what Hilda, Robbie, and Steven have been working on.

STANDARD 4:
TOOLS FOR ENHANCING DISCOURSE

The teacher of mathematics, in order to enhance discourse, should encourage and accept the use of—

♦ *computers, calculators, and other technology;*

♦ *concrete materials used as models;*

♦ *pictures, diagrams, tables, and graphs;*

♦ *invented and conventional terms and symbols;*

♦ *metaphors, analogies, and stories;*

♦ *written hypotheses, explanations, and arguments;*

♦ *oral presentations and dramatizations.*

Elaboration

In order to establish a discourse that is focused on exploring mathematical ideas, not just on reporting correct answers, the means of mathematical communication and approaches to mathematical reasoning must be broad and varied. Teachers must value and encourage the use of a variety of tools rather than placing excessive emphasis on conventional mathematical symbols. Various means for communicating about mathematics should be accepted, including drawings, diagrams, invented symbols, and analogies. The teacher should introduce conventional notation at points when doing so can further the work or the discourse at hand. Teachers should also help students learn to use calculators, computers, and other technological devices as tools for mathematical discourse. Given the range of mathematical tools available, teachers should often allow and encourage students to select the means they find most useful for working on or discussing a particular mathematical problem. At other times, in order to develop students' repertoire of mathematical tools, teachers may specify the means students are to use.

Vignettes

The teacher expects not just answers, *but also* reasons.

4.1 Mr. Johnson has presented his first-grade class with several pairs of numbers and asked them to decide which number is greater and to justify their responses. He has also been encouraging them to find ways to write these comparisons.

The child uses objects to communicate and justify his answer.

Ben: I think 5 is greater than 3 because (he walks to the board and sticks five magnets up and then carefully sticks three magnets in another row).

Mr. Johnson asks whether that makes sense to other people. The children nod. He asks if anyone wants to show how they would write this down.

The teacher encourages students to use symbols to represent and communicate about ideas.

Kevin, up at the board, writes:

$$5 \rightarrow 3$$

Next, Betsy writes:

5　(3)

Mr. Johnson: Can you explain what you were thinking? Kevin?

Kevin explains that his arrow shows that 5 is more than 3 because the bigger number "can point at" the smaller one.

Mr. Johnson asks Betsy to explain hers, and she says that she thinks you should just circle the smaller one.

Mr. Johnson: What if the two numbers you were comparing were 6 and 6? What would you do? How would you write that?

Several seconds pass. Ruth shoots her hand in the air. Several others also have their hands up.

Ruth: You could draw an arrow to both of them.

Annie: You could circle both of them because they are the same.

Jimmy: You shouldn't mark *either* one, either way. They are not greater or less. They are the same.

Mr. Johnson nods at their suggestions. He writes an equals sign (=) on the board and explains that this is a symbol that people have invented for the ideas the children have been talking about.

Rashida: That's like what Ruth said.

Ruth beams and Annie calls out: It's like mine too.

The teacher accepts more than one way of representing the idea with symbols; both are nonstandard but sensible.

The teacher poses a challenge that requires students to invent a means of recording an idea.

The teacher gives students time to think before responding. He doesn't repeat the question or call on children; he is silent.

The teacher connects the students' approaches and reasoning to the conventional notation. Because the students have thought about what it means for two numbers to be equal, they are ready to learn how that is conventionally represented. In this case, the notation follows the development of the concept in a meaningful context.

4.2 Mrs. Martinez and Mr. Golden, who have teamed up to teach eighth grade this year, have divided their students into groups of four. The teachers have challenged them to show why the text says that division by zero is "undefined." The teachers want their students to know why "you can't divide by zero." Usually that is all that students have learned. Once the students figure out why division by zero is undefined, they are to prepare something that they could use to justify their explanation to the rest of the class.

The teachers have posed a task that requires students to communicate about mathematics—in pairs as well as in the whole group—and that lends itself to a variety of tools.

Mrs. Martinez suggests that the calculator may be a useful tool for this problem. "Making up some kind of story problem for a situation that involves division might be helpful for others," adds Mr. Golden. The two teachers have arranged their large classroom so that calculators, graph paper, Unifix cubes and base ten blocks, felt-tip markers and blank overhead transparencies, rulers, and other materials are out where students can freely use them. This facilitates the use of alternative tools. Students are encouraged and expected to make decisions about which tool to use. Several students are, in fact, preparing overheads to display their conclusions about division by zero. Others are excitedly punching calculator buttons.

The teachers suggest particular tools in order to stimulate students to make choices about what might help them work on the problem.

"The answer keeps getting larger and larger!" exclaim a pair of girls as they watch the results obtained by successively dividing 4 by smaller and smaller divisors with the calculator. "Why is that important?" asks Mrs. Martinez as she watches over one girl's shoulder. "Well, because each

The teacher suggests another tool—a graph—that might help the students make mathematical connections, examine the pattern they are seeing, and present their work to the rest of the group.

of the numbers we are dividing by is getting closer and closer to zero but isn't zero." "Maybe you could make a graph to show what you are finding," suggests Mrs. Martinez.

Mr. Golden finds two students slouching sullenly in their chairs behind the room divider. "We don't understand what to do," grumbles one. Sitting down next to them, Mr. Golden begins, "Let's see if I can help. You are trying to figure out what the special problem is in trying to divide by zero. Maybe you can use some things you already know about division. How do you know that 8 ÷ 2 is 4? How could you prove that if someone challenged your answer?" The students look at him disbelievingly. He waits. Then one says, "Well, I'd just say that 4 times 2 is 8 so 8 divided by 2 *has* to be 4." "Can that help you at all with this problem?" asks Mr. Golden. He stands up. The two students look at one another and then, sitting up a bit, begin talking. "Well, that doesn't *work* if you take 8 ÷ 0," Mr. Golden hears one say as he walks away.

Summary: Discourse

Because the discourse of the mathematics class reflects messages about what it means to know mathematics, what makes something true or reasonable, and what doing mathematics entails, it is central to both *what* students learn about mathematics as well as *how* they learn it. Therefore, the discourse of the mathematics class should be founded on mathematical ways of knowing and ways of communicating. The nature of the activity and talk in the classroom shapes each student's opportunities to learn about particular topics as well as to develop their abilities to reason and communicate about those topics. Students' dispositions toward mathematics are also fundamentally influenced by the experiences they have with mathematical activity. Although teachers may seem quieter at times, the teacher is nevertheless central in fostering worthwhile mathematical discourse within the classroom community. Teachers' skills in developing and integrating the tasks and discourse in ways that promote students' learning depend on the construction and maintenance of a learning *environment* that supports and grows out of these kinds of thinking and activity. It is to this issue that we now turn.

In this situation, the teacher chooses to point the students directly to a means of working on and discussing the problem that plays off of important mathematical connections.

Having introduced this idea as a tool for working on the problem, the teacher leaves the students to use it on their own.

ENVIRONMENT

The mathematics teacher is responsible for creating an intellectual environment in which serious engagement in mathematical thinking is the norm, for the environment of the classroom is foundational to what students learn. More than just a physical setting with desks, bulletin boards, and posters, the classroom environment forms a hidden curriculum with messages about what counts in learning and doing mathematics: Neatness? Speed? Accuracy? Listening well? Being able to justify a solution? Working independently? If we want students to learn to make conjectures, experiment with alternative approaches to solving problems, and construct and respond to others' mathematical arguments, then creating an environment that fosters these kinds of activities is essential.

STANDARD 5:
LEARNING ENVIRONMENT

The teacher of mathematics should create a learning environment that fosters the development of each student's mathematical power by—

♦ **providing and structuring the time necessary to explore sound mathematics and grapple with significant ideas and problems;**

♦ **using the physical space and materials in ways that facilitate students' learning of mathematics;**

♦ **providing a context that encourages the development of mathematical skill and proficiency;**

♦ **respecting and valuing students' ideas, ways of thinking, and mathematical dispositions;**

and by consistently expecting and encouraging students to—

♦ **work independently or collaboratively to make sense of mathematics;**

♦ **take intellectual risks by raising questions and formulating conjectures;**

♦ **display a sense of mathematical competence by validating and supporting ideas with mathematical argument.**

Elaboration

This standard focuses on key dimensions of a learning environment in which serious mathematical thinking can take place: a genuine respect for others' ideas, a valuing of reason and sense-making, pacing and timing that allow students to puzzle and to think, and the forging of a social and intellectual community. Such a learning environment should help all students believe in themselves as successful mathematical thinkers.

What teachers convey about the value and sense of students' ideas affects students' mathematical dispositions in the classroom. Students are more likely to take risks in proposing their conjectures, strategies, and solutions in an environment in which the teacher respects students' ideas, whether conventional or nonstandard, whether valid or invalid. Teachers convey this kind of respect by probing students' thinking, by showing interest in understanding students' approaches and ideas, and by refraining from ridiculing students. Furthermore, and equally important, teachers must teach students to respect and be interested in one another's ideas.

Demonstrating respect for students' ideas does not imply, however, that teachers or students accept all ideas as reasonable or valid. The purpose of valuing students' ideas and ways of thinking is not just to make students feel good but to foster the development of their understanding of, and power with, mathematics. Therefore, the central focus of the classroom environment must be on sense-making. Mathematical concepts and procedures—indeed, mathematical skills—are central to making sense of mathematics and to reasoning mathematically. Teachers should consistently expect students to explain their ideas, to justify their solutions, and to persevere when they are stuck. Teachers must also help students learn to expect and ask for justifications and explanations from one another. Teachers' own explanations must similarly focus on underlying meanings; something a teacher says is not true simply because he or she "said so."

Emphasizing reasoning and justification implies that students should be encouraged and expected to question one another's ideas and to explain and support their own ideas in the face of others' challenges. Teachers must help students learn how to do this: Students need to learn how to question another's conjecture or solution with respect for that person's thinking and knowledge. They also need to learn how to justify their own claims without becoming hostile or defensive.

Serious mathematical thinking takes time as well as intellectual courage and skills. A learning environment that supports problem solving must allow time for students to puzzle, to be stuck, to try alternative approaches, and to confer with one another and with the teacher. Furthermore, for many worthwhile mathematical tasks, tasks that require reasoning and problem solving, the speed, pace, and quantity of students' work are inappropriate criteria for "doing well." Too often, students have developed the idea that if they cannot answer a mathematical question almost immediately, then they might as well give up. Teachers must encourage and expect students to persevere, to take the time to figure things out. In discussions, the teacher must allow time for students to respond to questions and must also expect students to give one another time to think, without bursting in, frantically waving hands, or showing impatience.

Students' learning of mathematics is enhanced in a learning environment that is built as a community of people collaborating to make sense of mathematical ideas. It is a key function of the teacher to develop and nurture students' abilities to learn with and from others—to clarify definitions and terms to one another, consider one another's ideas and solutions, and argue together about the validity of alternative approaches and answers. Classroom structures that can encourage and support this collaboration are varied: students may at times work independently, conferring with others as necessary; at other times students may work in pairs or in small groups. Whole-class discussions are yet another profitable format. No single arrangement will work at all times; teachers should use these arrangements flexibly to pursue their goals.

Vignettes

5.1 A class of primary students has been working on problems that involve separating or dividing. The teacher, Laurie Morgan, is trying to give them some early experience with multiplicative situations at the same time that she provides them with contexts for deepening their knowledge of and skill with addition and subtraction. These students can add and subtract, but their understanding of multiplication and division is still quite informal. They have begun to develop some understanding of fractions, connected to their ideas about division. They have not yet learned any conventional procedures for dividing.

The teacher has selected this problem because it is likely to elicit alternative representations and solution strategies as well as different answers. It will also help the students develop their ideas about division, fractions, and the connections between them.

Today Mrs. Morgan has given them the following problem:

> If we make 49 sandwiches for our picnic, how many can each child have?

After they have worked for about twenty minutes, first alone and then in small groups, Mrs. Morgan asks if the children are ready to discuss the problem in the whole group. Most, looking up when she asks, nod. She asks who would like to begin.

Two girls go to the overhead projector. They write:

$$\begin{array}{r} 49 \\ -28 \\ \hline 22 \end{array}$$

The teacher allows time for the children to develop their solutions independently, with a few others, and then in the whole group. By asking who would like to share their solution, she encourages the students to take intellectual risks.

One explains, "There are twenty-eight kids in our class, and so if we pass out one sandwich to each child, we will have twenty-two sandwiches left, and that's not enough for each of us, so there'll be leftovers."

Students expect to have to justify their solutions, not just give answers.

The teacher and students are quiet for a moment, thinking about this. Then Mrs. Morgan looks over the group and asks if anyone has a comment or a question about this solution.

The teacher solicits other students' comments about the girls' solution without labeling it right or wrong. She expects the students, as members of a learning community, to decide if an idea makes sense mathematically.

One boy says that he thinks their solution makes sense, but that "nine minus eight is one, not two, so it should be twenty-one, not twenty-two." He demonstrates by pointing at the number line above the chalkboard. Starting at nine, he counts back eight using a pointer. The two girls ponder this for a moment. The class is quiet. Then one says, "We revise that. Nine minus eight *is* one." Mrs. Morgan is listening closely, but does not jump into the interchange.

Students respectfully question one another's ideas. The girls "revise" their solution because they have been convinced by the boy's explanation. There is no sense here that being wrong is shameful.

Another child remarks that he had the same solution as they did—one sandwich.

"Frankie?" asks Mrs. Morgan, after pausing for a moment to look over the students. She remembers noticing his approach during the small-group time. Frankie announces, "I think we can give each child *more* than one sandwich. Look!" He proceeds to draw twenty-one rectangles on the chalkboard. "These are the leftover sandwiches," he explains. "I can cut fourteen of them in half and that will give us twenty-eight half-sandwiches, so everyone can get another half."

The students work together to solve the problem. Sometimes they build on the solutions offered by classmates. The teacher gathers insights about students through close listening and observation. At times, she takes responsibility for pushing students' thinking along.

"I agree with Frankie," says another child. "Each child can have one and a half sandwiches."

"Do you have any leftovers?" asks the teacher.

The teacher expects the students to reason mathematically.

"There are still seven sandwiches left over," says Frankie.

"What do the rest of you think about that?" inquires the teacher.

Several children give explanations in support of Frankie's solution. "I think that does make sense," says one girl, "but I had another solution. I think the answer is one plus one-half plus one-fourth."

Students seem willing to take risks by bringing up different ideas.

"I don't understand," Mrs. Morgan says. "Could you show what you mean?"

The teacher expects students to clarify and justify their ideas.

5.2 Mr. Cohen's class of high school students is working in small groups on projects that involve collecting, organizing, and interpreting data. Before beginning, they had a discussion about different possibilities for their projects. They decided on questions that they would like to pursue, such as finding the average number of hours per week that high school students work.

The teacher has posed a task that gives the students an opportunity to develop their understanding of "sample" and "population," as well as to build their ability to use statistics to reason about real-world situations.

One group, having read an article in the newspaper on changes in the popularity of first names in the last sixty-five years, has decided to investigate the most common boy's and girl's first names among students their age in the city. Is John still the most common boy's name, as it was in 1925, 1950, and 1975? They are curious about what is happening with girls' names, since the popular names seem to change more often. Because theirs is an ethnically diverse community, they also wonder how that affects the pattern of names.

Mr. Cohen works his way around to the different groups, listening, making suggestions, and verifying that the group members are listening and working together. The group that is working on the names study has decided to sample the high school population in the city and is discussing the best way to go about this.

John: Let's choose three of the high schools and then write to them and ask for a list of the students enrolled in the school. We can take our sample from those lists at random.

Jenny: But how will we pick the three schools? And why is three a good number to pick?

John: It seemed like enough out of all the schools in this city *if* we were careful to include one of the schools that has more kids from different backgrounds, because we want to make sure our sample has lots of different kinds of names, just like there are around here.

Anna: I think we should try to figure out about how many high school kids there are in the whole city and then pick a size for our sample based on that.

John (nodding): I guess that makes sense. How are we going to figure that out, though?

Maria: And then how big would our sample have to be to be big enough? We want to be pretty sure that our sample tells us something about all the kids in high school here.

Mr. Cohen is standing by the group. He says that their discussion so far is productive, that they are dealing with some important questions for their project. He suggests a source that might help them think about the question of how many students they need to have in their sample. He also tells them that the administration office would have a list of all the high schools and how many students attend each of them. He asks if they would like him to call and ask for that list. They say that they would. He asks what they are going to do once they get the list.

5.3 Mr. Davies, who teaches seventh grade, wants to begin encouraging a more collaborative environment in his classroom. However, he finds that his students seem anxious about having the right answer. They do not reflect on whether or not their answers make sense—as long as they match the answer key. They do like to work in groups, but it seems that the attraction is primarily social.

For example, one day when they were working on a set of ratio and proportion exercises and story problems, he told them to form small groups to discuss the work. The groups were loud, and in each group someone seemed to dominate while others sat by passively. When their answers differed, they usually erased their answers to match one another's. Typically the answer considered "right" was the answer given by Julio or Evie—or whomever they considered smartest in math.

The task itself promotes collaboration. It is not clear how to proceed, and everyone's contributions are needed in order to come up with an approach.

The teacher is providing time for students to grapple with the problems.

These students seem to have developed the disposition to question one another, and they respond to one another's queries as a matter of course.

Helping a group of students develop a sense of community is facilitated by having some sense of shared purpose.

The students seem accustomed to thinking through problems together.

The teacher monitors how the group is proceeding. He also offers resources where appropriate.

It can be difficult to get students who are used to working alone to understand or appreciate collaborative work. Simply moving students into small groups is not likely to prepare them adequately to work together well. The task also makes a difference. Here, these exercises may not spur discussion in the way that another more problematic or intellectually challenging task might.

Discouraged, Mr. Davies consults with the teacher next door, who offers to come observe during her gym period the next day. As they talk, she suggests that the tasks he is using in the groups may be contributing to the problems. Together they design a couple of tasks dealing with ratio and proportion that they think might get the students thinking more and also provoke better discussion in the groups. When she observes him, she also notices that he frequently praises students for having the right answer and prods those who are not doing their "own work." She points this out to him gently, confessing that this gives her a hard time, too. He says, ruefully, "I guess we have some habits that give the kids some mixed messages at times."

Summary: Environment

The learning environment is a key element in fostering the goals of the *Curriculum and Evaluation Standards for School Mathematics*. Creating an environment that supports and encourages mathematical reasoning and fosters all students' competence with, and disposition toward, mathematics should be one of the teacher's central concerns. The nature of this learning environment is shaped by the kinds of mathematical tasks and discourse in which students engage. Teachers' skills in developing and integrating the tasks, discourse, and environment in ways that promote students' learning are enhanced through thoughtful analysis of their instruction, which is the focus of the last section of the standards for teaching.

Teachers can help one another focus on and enhance the learning environment.

Tasks affect both the environment and the discourse. Different tasks require different types of instructional strategies.

Teachers give off many cues about what is valued. When teachers are trying to change their approach, sometimes the messages students receive can get mixed.

♦   ♦   ♦   ♦   ♦   ♦   ♦   ♦

ANALYSIS

A central question for which teachers must be responsible is, "How well are the tasks, discourse, and environment working to foster the development of students' mathematical literacy and power?"

Trying to understand as much as possible about the effects of the mathematics classroom on each student is essential to good teaching. Teachers must monitor classroom life using a variety of strategies and focusing on a broad array of dimensions of mathematical competence, as outlined in the *Curriculum and Evaluation Standards for School Mathematics*. What do students seem to understand well, what only partially? What connections do they seem to be making? What mathematical dispositions do they seem to be developing? How does the group work together as a learning community making sense of mathematics? What teachers learn from this should be a primary source of information for planning and improving instruction in both the short and the long term.

STANDARD 6:
ANALYSIS OF TEACHING AND LEARNING

The teacher of mathematics should engage in ongoing analysis of teaching and learning by—

◆ *observing, listening to, and gathering other information about students to assess what they are learning;*

◆ *examining effects of the tasks, discourse, and learning environment on students' mathematical knowledge, skills, and dispositions;*

in order to—

◆ *ensure that every student is learning sound and significant mathematics and is developing a positive disposition toward mathematics;*

◆ *challenge and extend students' ideas;*

◆ *adapt or change activities while teaching;*

◆ *make plans, both short- and long-range;*

◆ *describe and comment on each student's learning to parents and administrators, as well as to the students themselves.*

Elaboration

Assessment of students and analysis of instruction are fundamentally interconnected. Mathematics teachers should monitor students' learning on an ongoing basis in order to assess and adjust their teaching. Observing and listening to students during class can help teachers, on the spot, tailor their questions or tasks to provoke and extend students' thinking and understanding. Teachers must also use information about what students are understanding to revise and adapt their short- and long-range plans: for the tasks they select and for the approaches they choose to orchestrate the classroom discourse. Similarly, students' understandings and dispositions should guide teachers in shaping and reshaping the learning environment of the classroom. Additionally, teachers have the responsibility of describing and commenting on students' learning to administrators, to parents, and to the students themselves.

Students' mathematical power depends on a varied set of understandings, skills, and dispositions. Teachers must attend to the broad array of dimensions that contribute to students' mathematical competence as outlined in the *Curriculum and Evaluation Standards for School Mathematics*. They should assess students' understandings of concepts and procedures, including the connections they make among various concepts and procedures. Teachers must also assess the development of students' ability to reason mathematically—to make conjectures, to justify and revise claims on the basis of mathematical evidence, and to analyze and solve problems. Students' dispositions toward mathematics—their confidence, interest, enjoyment, and perseverance—are yet another key dimension that teachers should monitor.

Paper-and-pencil tests, although one useful medium for judging some aspects of students' mathematical knowledge, cannot suffice to provide teachers with the insights they need about their students' understandings in order to make instruction as effectively responsive as possible. Teachers need information gathered in a variety of ways and using a range of sources. Observing students participating in a small-group discussion may contribute valuable insights related to their abilities to communicate mathematically. Interviews with individual students will complement that information and also provide information about students' conceptual and procedural understanding. Students' journals are yet another source that can help teachers appraise their students' development. Teachers can also learn a great deal from closely watching and listening to students during whole-group discussions.

As they monitor students' understandings of, and dispositions toward, mathematics, teachers should ask themselves questions about the nature of the learning environment they have created, of the tasks they have been using, and of the kind of discourse they have been fostering. They should seek to understand the links between these and what is happening with their students. If, for example, students are having trouble understanding inverse functions, is it because of the kinds of tasks in which they have been engaged? Is it related to the ways in which the group has explored and discussed ideas about functions and their inverses? Although it may be that the students lack prerequisite understandings, it could also be that this is a difficult piece of mathematics or that the teacher needs to consider alternative ways to help students "unpack" the ideas. Or, if students quickly give up when a direct route for solving a problem is not apparent, teachers must consider how the experiences that students have been having and the environment in which they have been working may not have helped them to develop the perseverance and confidence they need. Teachers need to analyze continually what they are seeing and hearing and explore alternative interpretations of that information. They need to consider what such insights suggest about how the environment, tasks, and discourse could be enhanced, revised, or adapted in order to help students learn.

Vignettes

6.1 Some teachers begin to change by allocating one day a week to "different" mathematics activities. Although the *Curriculum and Evaluation Standards* makes clear that the goal is for problem solving, reasoning, and communication to be interwoven throughout the curriculum, teachers must experiment with alternative approaches to changing their practice. This "one day a week" strategy is one such approach—not the goal, but for some teachers, a viable first step.

Ms. Levesque has been having students working in groups of four on Fridays, solving nonroutine problems. Last week, she had them work on the handshake problem. (If ten people are at a party and everyone shakes everyone else's hand exactly once, how many handshakes take place?) Things seem to be going quite well. The students appear to enjoy these Friday sessions and she looks forward to them herself.

The teacher gets information about her students from informal as well as formal sources. Students' conversations often give her clues about how they are feeling about mathematics or mathematics class.

She notices, however, that when she listens to students talking among themselves before class, they still groan about the word problems on their daily homework. Many students leave this part of their work unfinished because, they say, the problems are too hard.

As Ms. Levesque compares what she is learning about her students with her goals for them, she is troubled, because she wants her students to feel confident about solving mathematical problems and to stick to them even when the problems are hard.

She thinks about those Friday sessions. Why aren't they fostering these dispositions toward mathematical problem solving? Ms. Levesque wonders whether perhaps these special sessions seem to the students to be separate from "real math." It is, after all, just one day a week—and what they do on the other four days is quite different in spirit and in content.

Ms. Levesque decides to try working on word problems together for part of the period every day for a while to see if that makes a difference. She will try having them discuss the problems, examining different approaches and solutions, instead of just going over the answers together. In addition, the students will keep a journal or notebook in which to record strategies and reflections. When she talks to her department head about this over lunch, her department head says that she has had a similar concern with her classes and that she, too, will try Ms. Levesque's plan and they can compare notes after a few weeks.

The teacher analyzes what she has been doing—the tasks she has been using and the environment and discourse she has created around them—and comes up with an alternative plan to try, based on her analysis of the situation.

Talking and collaborating with colleagues can enhance the analysis and the process of improving instruction.

6.2 The second graders have just finished working on addition and subtraction with regrouping. On a written test, many of them "forget" to regroup when they need to in subtraction. Instead, they do this:

$$\begin{array}{r} 50 \\ -38 \\ \hline 28 \end{array}$$

The teacher gathers information about what students have learned.

The teacher, Mr. Lewis, thinks they are being careless. He feels a little annoyed because this is something on which he has spent a lot of time. He decides, though, that he should sit down with the children one by one for a few minutes and have them talk through a couple of the problems and how they solved them. He thinks he may be able to tell what they are doing wrong this way.

He chooses a couple of problems from the test and asks the children to justify their answers using bundles of Popsicle sticks. He discovers that most of them are not connecting the work they did in class with manipulatives to these written problems. When they have the Popsicle

Instead of relying on his assumptions to explain what he has found, the teacher decides to gather some more information that might help him understand what has gone wrong. He assumes that there may be reasons for the children's performance that go beyond carelessness.

The teacher develops a reasonable strategy for gathering some additional information about the students' understanding.

sticks, they find that their answers don't make sense, and they revise them to match what they do with the sticks.

Mr. Lewis had assumed that if they "saw" the concepts by actually touching the objects, they would understand. He now thinks that maybe he didn't do enough to help them build the links between the concrete model and the algorithm. He starts wondering what he could do to help them make that connection better. He hypothesizes that maybe they know *how* to regroup but may not understand *why* or *when* regrouping is necessary. He decides to make up a worksheet with examples where regrouping is necessary and some where it is not and have the children discuss whether or not they would have to regroup in each case and how they know that.

6.3 Ms. Lundgren has been trying to change her approach to teaching mathematics so that students are learning to reason and communicate about mathematics, to make sense of mathematical ideas, and to make connections. She believes she has been successful in moving the discourse of her classroom away from a focus on right answers and the teacher as authority.

Although she finds it difficult, she has also been devising better mathematical tasks, she thinks. With the help of the other fifth-grade teacher, Ms. Lundgren has also come up with some ways of keeping track of what students are learning. Today she is meeting with parents to go over their children's report cards, and she has decided to draw on her new records for these conferences.

When she is meeting with Mrs. Byers, Stacy's mother, Ms. Lundgren wants to show her how Stacy is making connections in division. Looking at her card on Stacy, Ms. Lundgren tells the mother that Stacy was able to explain how, for 28 ÷ 8, 3 r 4 was the same answer as 3.5 (a quotient obtained on the calculator) but also how the two answers differed. Ms. Lundgren, having made a note of it, opened Stacy's mathematics journal to the page where Stacy had worked this out. Then, referring to the index card again, Ms. Lundgren shows Stacy's mother all the ways that Stacy found to represent $8 \div \frac{1}{2}$ in her journal. Because she also wants to talk with Mrs. Byers about Stacy's disposition toward mathematics, Ms. Lundgren refers to a chart she is keeping on her students' mathematical attitudes. With this chart, she has periodically made notes to herself. She has also had her colleague next door come in and observe once a month and make notes on the chart for her. Mrs. Byers finds all these specific examples very useful and comments that she thinks that what Ms. Lundgren is trying to do in math is great and she wishes she had had a mathematics class like this when she was in school.

6.4 Ms. Weissmann has been audiotaping her mathematics classes each day this year. She listens to as much of each tape as possible while she plans for the next day's class. In listening to herself and to the students, she begins to notice a pattern.

On the one hand, the girls are very quiet and speak softly and say "I don't know" at least as often as they say anything. The boys, on the other hand, are loud, and she hears herself calling them by name a lot. They participate actively in the mathematics discussions as well as in their

The teacher analyzes what he finds from talking individually with the students. He reflects on how he worked with the class on this topic and conjectures that his approach had some flaws. He begins a search for how he can revise what he was doing.

The teacher knows that she must find some ways of documenting and assessing what students are learning, especially in view of her new goals for them. She finds it helpful to work with a colleague.

The teacher wants the parent to understand both what her child is doing and what is being held as important in her mathematics class.

Because it enables her to give the parent specific examples, her system of cards as an index to the children's journals helps her to do both.

The teacher got this idea from the NCTM Curriculum and Evaluation Standards for School Mathematics,(pp. 235, 236). She and her colleague found several ideas there for assessing and keeping track of students' learning.

The teacher has a systematic way of collecting and analyzing information about her own teaching.

own little games and fooling around. She begins tallying the frequency with which she calls on boys and on girls. She also begins a chart for *what* the boys and girls each contribute to class discussions, not just how often.

At the same time, Ms. Weissmann gets a couple of books from the library, both centered on discourse and on women's patterns of interaction in different settings. She decides to make this a project for herself: to improve the balance of kinds and frequency of participation among boys and girls in the class discussion. She also plans to be alert if there are other such patterns underlying the boy-girl split.

Summary: Analysis

Analysis of instruction recognizes the intimate relationship between teaching and assessment. To improve their mathematics instruction, teachers must constantly analyze what they and their students are doing and how that is affecting what the students are learning. Using a variety of strategies, teachers must continuously monitor students' capacity and inclination to analyze situations, frame and solve problems, and make sense of mathematical concepts and procedures. Teachers should use such information about students to assess not just how students are doing, but also to appraise how well the tasks, discourse, and environment are working together to foster students' mathematical power and to adapt their instruction in response.

This pattern is not uncommon, but it is troubling to this teacher, who has always been interested in, and relatively successful with, mathematics. She also is convinced that things do not have to be like this.

The teacher selects some simple ways of maintaining a record of what is going on in her class.

The teacher's "project" helps her to focus on an issue that is of great importance to her.

STANDARDS FOR THE EVALUATION OF THE TEACHING OF MATHEMATICS

OVERVIEW

This section presents eight standards for the evaluation of the teaching of mathematics organized under two categories:

The Process of Evaluation

1. The Evaluation Cycle

2. Teachers as Participants in Evaluation

3. Sources of Information

The Foci of Evaluation

4. Mathematical Concepts, Procedures, and Connections

5. Mathematics as Problem Solving, Reasoning, and Communication

6. Mathematical Disposition

7. Assessing Students' Mathematical Understanding

8. Learning Environment

INTRODUCTION

Efforts to improve the teaching of mathematics are necessarily a function of what good mathematics teaching is considered to be. Deciding which aspects of teaching need to be improved requires both information about the teaching process and a framework that suggests what we value. The previous section presents a vision for teaching mathematics based on the *Curriculum and Evaluation Standards for School Mathematics*. The standards in this section are intended to help teachers attain that vision by emphasizing the role that evaluation can play in teachers' professional development. In keeping with the notion that assessment is a process of gathering and interpreting information, these standards focus on how and what information should be gathered and analyzed to help teachers improve their teaching.

The assessment process described in the following standards can be used by a teacher engaged in a process of self-analysis and personal growth or by a teacher working in concert with colleagues or supervisors in an effort to improve instruction. Each standard serves as a statement about what should be observed regardless of who is doing the observing. Further, the standards can be useful in evaluating teachers with a wide range of teaching experience and expertise. They provide foci to be considered by all who teach mathematics.

Evaluation comes in many forms. Teachers improve their own teaching by reflecting on and analyzing previous lessons. Evaluation is also an activity involving peers, one supporting the other in trying to improve the quality of instruction. The vignettes in this section illustrate both forms of evaluation.

Teachers are sometimes evaluated because of district or state mandates to make career-ladder decisions or to make judgments about a teacher's competence. It is our position that such evaluations should also adhere to the following standards particularly with respect to the teacher's participation in the process and in defining the foci of the evaluation.

Teachers who strive to improve their instruction will take risks by experimenting with instructional approaches that are either new to them or that they have not yet mastered. The assessment process should not restrict a teacher's willingness to take those risks, nor should it be permitted to interfere with instruction in any other way. Teachers need freedom and support to develop professionally. For a teacher engaged in experimentation, self-analysis or conferring with a colleague may be more appropriate than evaluation by an administrator.

It is imperative that the teaching of mathematics enable every student to become mathematically powerful and that we increase the participation of all students in the study of mathematics. By "every student" we mean specifically—

♦ students who have been denied access in any way to educational opportunities as well as those who have not;

♦ students who are African American, Hispanic, American Indian, and other minorities as well as those who are considered to be a part of the majority;

♦ students who are female as well as those who are male;

♦ students who have not been successful in school and in mathematics as well as those who have been successful.

Thus, an important consideration in evaluating the teaching of mathematics is whether the mathematical needs of every student are being addressed.

ASSUMPTIONS

The standards in this section are based on the following four assumptions:

1. *The goal of evaluating the teaching of mathematics is to improve teaching and enhance professional growth.* The teacher is the key to high-quality mathematics education for students. It is the teacher who makes decisions about curriculum and teaching methods to maximize student learning. Professional development extends and expands teachers' abilities to make good decisions by giving them access to a deeper understanding of mathematics, a greater knowledge about students' learning of mathematics, a greater repertoire of teaching strategies, and the ability to match their repertoire to the needs of all students. Evaluation helps identify the teacher's needs so that appropriate professional development experiences can be provided.

2. *All teachers can improve their teaching of mathematics.* These standards are intended for all teachers. Whether beginning or experienced, all teachers can find some aspect of their teaching that can be improved by considering one or more of the standards. Though experienced teachers might be more adept at self-analysis, beginning or struggling teachers can also reflect on the standards and arrive at some conclusions on how their teaching can be improved.

3. *What teachers learn from the evaluation process is related to how the evaluation is conducted.* The primary emphasis of the standards is

on improvement through self-analysis and working in a collegial and supportive environment with peers, supervisors, and administrators. Evaluations can be conducted in many ways, and many sources of information should be used. When written evaluation reports for a teacher's personnel file are produced, a spirit of sensitivity, mutual respect, and concern for professional growth as the primary purpose of evaluation are especially important.

4. *Because teaching is complex, the evaluation of teaching is complex. Simplistic evaluation processes will not help teachers realize the vision of teaching mathematics described in these standards.* Teaching is a function of many activities including listening, informing, stimulating, challenging, and motivating. These and other activities should be done in ways that are responsive to students and take advantage of their knowledge and disposition to do mathematics. A particularly sensitive issue related to the complexity of evaluating teaching is whether and how information about students' understanding of, and disposition to do, mathematics should be considered. It seems only reasonable that students' progress should provide *a* source of information about teaching. However, student learning of, and disposition to do, mathematics should not be the only sources of information. It follows that any evaluation process that intends to help teachers achieve the vision of teaching mathematics suggested in this volume should consider numerous factors and circumstances and should have a longitudinal and cyclical perspective.

The next section contains three standards that describe the evaluation process and its contribution to a teacher's professional development. The subsequent section contains five standards that provide the foci for evaluating the teaching of mathematics.

THE PROCESS OF EVALUATION

The evaluation process should generate information about teaching and provide an analysis of that information which then leads to rich and appropriate professional development experiences. Such professional development should lead to the improvement of instruction. The standards in this section address the evaluation process and its connection to professional development.

The complexity of teaching requires an evaluation process based on information from a variety of sources and a variety of teaching situations. Teachers may demonstrate more strengths in one context than in another. A fair and valid evaluation process should collect enough data from a variety of contexts to allow an accurate description of the teacher's abilities.

Improving the teaching of mathematics depends on what the teacher knows and does. The evaluation process can reveal areas of instruction that are not consistent with the desired vision of teaching mathematics, but only the teacher can take steps toward realizing the vision. Accordingly, the teacher is the key element in the assessment process and, consequently, should be fully involved in determining which aspects of teaching should be the focus of individual professional development.

STANDARD 1:
THE EVALUATION CYCLE

The evaluation of the teaching of mathematics should be a cyclical process involving—

♦ *the periodic collection and analysis of information about an individual's teaching of mathematics;*

♦ *professional development based on the analysis of teaching;*

♦ *the improvement of teaching as a consequence of the professional development.*

Elaboration

Evaluation is the vehicle that connects a teacher's current teaching with the professional development necessary to enable that teacher to improve the teaching of mathematics. The evaluation process begins by collecting data representative of the teacher's current practice. The collected data is then analyzed with respect to what is valued in the teaching of mathematics, such as the vision of teaching presented in the first section of this volume. Aspects of instruction that are deemed consistent with what is valued should be identified as well as those needing improvement. Although this analysis may result in a report for the teacher's personnel file, the more important outcome is the creation of a plan to help the teacher develop professionally. This plan should consist of instructional alternatives that have the potential for improving teaching as well as strategies for implementing these alternatives. Subsequent lessons are then observed and analyzed to determine whether improvement has been made; hence, the evaluation process is cyclical.

The cycle may require only a few minutes, as would be the case if a teacher thoughtfully reviews an algebra lesson taught during one period before teaching the lesson again during a later period, or it may require a year, if college coursework is recommended as a professional development activity. In most cases the length of the cycle would be between those two extremes. For example, a teacher may be trying to increase her repertoire of assessment techniques and is interested in determining the impact of the various techniques on student learning and disposition to do mathematics within a given grading period.

Too often the evaluation process involves only a supervisor making a single observation during an academic year. This process is limited in at least three ways. First, annual observations are much too infrequent to provide the basis for a comprehensive professional development plan. Second, evaluations by a single observer are too unreliable and ignore the wealth of expertise available from the teacher and the teacher's colleagues (see Standard 2). Third, evaluations based on a single source, such as a single classroom observation, are similarly unreliable and ignore other important sources of data that furnish additional information about teaching that would be useful for planning professional development (see Standard 3).

Professional development can take many forms (see the third section of this volume), including independent study, participation in in-service programs provided by the school, enrollment in college courses, discussions with colleagues, observations of colleagues, and attendance at professional meetings. Evidence of successful professional development should appear in subsequent teaching and be documented in future assessments.

The major goal for any evaluation of mathematics teaching should be to improve teaching and enhance professional growth. This emphasis would be a significant change from present evaluation practices in many school districts in which the goal is to provide documentation for personnel decisions or simply to comply with a requirement that all teachers have an assessment report added to their file according to some specified schedule. Although it may not be possible, or even desirable, to eliminate such reports from the evaluation process, it is critical that the primary emphasis be placed on the use of evaluation to furnish the basis for professional development activities aimed at improving the teaching of mathematics.

Vignettes

The school prepares mentor teachers to support the professional growth of young teachers and provides release time for their work.

The mentor teacher collects data on her colleague's teaching.

1.1 Before school begins, Jan Williams, an experienced fourth-grade teacher, is assigned to work as a mentor with Tom Burton, a first-year fifth-grade teacher at Valley Elementary School. Their classrooms are across the hall from each other, so Jan has many opportunities to observe Tom's class and confer with him about his professional development. On this day Jan observes Tom handing out worksheets to his class after reviewing the standard multiplication algorithm. She takes extensive notes that describe the students as well behaved but often uninvolved and passive. Tom tends not to ask many questions; Jan notes that of the fifteen questions she observed, all but one required a response of a number or a single word. Several students quickly finish their work and set it aside.

The mentor teacher observes students at work.

When Jan talks with the students, she notes that they have made a number of errors in doing the worksheets. They don't seem very interested in checking their work, however.

The teacher reflects on the lesson and discusses the lesson with the mentor teacher. The mentor teacher is helping him develop professionally.

The focus of the discussion is on the questioning techniques that a teacher can use.

That afternoon, Jan and Tom meet to discuss and analyze the information she has collected. She asks Tom to reflect on the lesson and give his evaluation of it. Tom recalls that several of his questions didn't seem to spark much discussion; Jan shares her observation that most of his questions could be answered by a number fact, a simple computation, or a single-word response. He says that he is so concerned with his own teaching activities that he neglects to focus on what the students are doing. He admits that often the students seem uninterested in the mathematics lessons but that he is hard pressed to figure out any alternatives. Besides, he explains, this particular lesson was a review lesson; he queries Jan on whether there are any better ways of conducting review lessons. Jan offers several suggestions on how he could rephrase his questions so that students would become more involved in class discussions and on various types of activities he could use to review the material more effectively. Tom decides that he will work on improving his skill in questioning students.

A goal is set for improving a teaching skill.

Together the teacher and the mentor identify management strategies to monitor students' work.

When discussing the part of the lesson involving seatwork, Tom recalls that some students' primary objective appears to be to finish the work ahead of the other students. Jan and Tom discuss ways of changing this attitude. Jan explains the differences between monitored practice, seatwork, and homework to help Tom plan how to structure class time. She reminds Tom that the teachers have been working with Claude Andrews, the principal and curriculum coordinator, to encourage the use of different teaching techniques.

In November, Claude makes his second visit to Tom's class after discussing Tom's progress with Jan. This time the class is studying the area of rectangles. The students are asked to draw rectangles that would have an area of 20 square units, each unit being a 1-inch square that the students have cut out of card stock. The principal observes that Tom monitors the students' work by looking at their drawings and quietly interacting with them as they work. This activity is followed by a discussion with different students sharing their work on the overhead. The following drawings are put on the overhead.

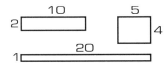

The principal notes that the teacher has improved his teaching by incorporating concrete materials and by improving the way he asks questions and interacts with the students. It is clear to the principal that the mentor has helped the teacher develop more effective ways of managing students and improving the students' attitudes toward mathematics and their willingness to work.

During the conference with Tom, Claude compliments him on implementing Jan's suggestions. Claude suggests that since the students have just reviewed perimeter, Tom incorporate problem solving involving both perimeter and area into the lesson by using questions like the following:

The evaluation cycle continues as a means of staff development.

The principal offers suggestions for improving the teaching of mathematics.

If the perimeter of a rectangle is 18 centimeters, what are the dimensions of the rectangle with the smallest area if the lengths of the sides are whole numbers?

If a rectangle has a length 6 times its width and has an area less than 50 square centimeters, what could be the dimensions of the rectangle, assuming the lengths of the sides are whole numbers?

The principal suggests that as a technique to enhance communication and involvement, Tom could have the students work in pairs. Tom thinks this is a good suggestion.

After discussing how to get students to work in pairs or in small groups, Claude lends Tom a reference book on cooperative learning and suggests that he observe Jan's class again, since she often uses cooperative learning groups when teaching mathematics. He schedules another set of observations with Tom in February.

References are provided for professional development.

Additional phases of the evaluation cycle are scheduled.

1.2 Three weeks before school begins, Mary Fisher examines two videotapes of her ninth-grade algebra class from last spring. As a seventeen-year veteran, she does a yearly review of goals and expectations before the new school year begins. The videotape helps her recall ending last year feeling vaguely disturbed by her inability to find classroom time to emphasize more problem-solving activities and discussions. She recognizes the need to find time to integrate student exploration and computer-based modeling into the required curriculum. Yet, she barely has enough time to cover the homework and present the new material.

The teacher assumes responsibility for periodic review and collecting information about her teaching.

The teacher senses that the problem involves the allocation of time within a given class period.

Self-analysis of the videotape reveals a startling observation regarding her allocation of class time. She notices that she spends almost twenty-five minutes, nearly the first half of the period, covering homework. More important, she observes that many of the students are off-task and passive while she does the problems on the board.

The teacher analyzes her teaching performance using a videotape.

Mary calls Delores Laco, a ninth-grade teacher and colleague. She asks Delores's opinion about changing their homework-review techniques this fall. Delores tells Mary about a recent journal article that she has read during the summer. The article offers various suggestions for reviewing

Professional development includes reading professional journals and collaborating with colleagues.

homework. Mary says that she will read the article. They agree to discuss alternatives for covering homework when they meet during the in-service days the last week in August.

The teacher changes her teaching practice as a consequence of her professional development activities.

Subsequently, Mary and Delores decide to try four different methods of covering homework at various times during the first quarter:

1. Have students keep their homework in a notebook that will be periodically reviewed.
2. Pair students to discuss their homework briefly at the beginning of the class period.
3. Give frequent short quizzes on the homework.
4. Write solutions to selected problems on a transparency and put these solutions on the overhead sometime during the class period.

Mary discusses these strategies with the students during the first week of school. In addition, she decides to start each class period with a problem-solving activity, moving homework to later in the period.

The log allows for a periodic collection of information for analysis.

In an effort to monitor her current classroom time spent on homework, Mary keeps a daily log of the amount of time spent going over homework in class. Delores suggests that they also keep track of the part of the period in which homework review takes place. In early October, Mary reviews her second-hour algebra class chart for the previous week.

	M	T	W	TH	F
Time started	9:18	9:40	9:33	9:20	No school
Total time	12 min.	5 min.	14 min.	9 min.	

The teacher reflects on her improvement of teaching as a consequence of her professional development activities.

Mary shares the results with Delores. Mary is very pleased that she has been able to reduce significantly the amount of time she has been spending on homework. Yet she has not detected any drop in student performance as a result of this new approach. To the contrary, the students seem to be more attentive when covering homework. In addition, she is pleased with her attempts to engage students in more problem-solving activities and discussions.

The teacher notes student difficulties and discusses them with his mathematics supervisor.

The supervisor notices a particular problem.

The teacher confers with the supervisor about the learning problem. The supervisor makes a suggestion for addressing the problem.

1.3 Jim Waseskuk is discussing his seventh-grade mathematics class with his mathematics supervisor, Ellen Davenport, as part of his quarterly evaluation. He tells her that the class is studying relationships between parallelograms and rectangles. Jim indicates that the class can identify properties of a given figure but that they are having difficulty making comparisons between parallelograms and rectangles. When Ellen observes the class, she notes that many students have difficulty with questions such as "How are the diagonals of a rectangle different from the diagonals of a parallelogram?"

During a planning period, Jim and Ellen discuss the problem. Jim expresses frustration at not being able to get the students to visualize the various properties of rectangles and parallelograms and, in particular, the diagonals of the figures. Ellen suggests using cardboard strips with brads at the corners to form a parallelogram that could be moved to form a rectangle. Ellen also suggests that elastic thread could be used to demonstrate how the diagonals change as the parallelogram becomes a rectangle. Jim likes the idea and thinks that he will have each of his students make a figure similar to the one Ellen has described.

The next day Jim describes to the students how to make the figures. He demonstrates how they should work using one he has made the night before.

The teacher follows through by constructing the figure and demonstrating it to students.

The students bring their constructed figures to class and explore the following motions with them.

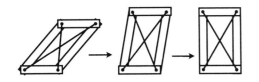

Jim is very pleased that the students are beginning to understand what happens when a figure with given sides is transformed from a parallelogram without right angles into a rectangle. During their explorations, they conjecture that the diagonals become congruent when the parallelogram becomes a rectangle but that other properties remain unchanged—for instance, that the diagonals of both parallelograms and rectangles bisect each other.

The teacher analyzes the lesson and notices the improvement in students' learning on the basis of the questions they can now answer correctly.

Ellen sees Jim at a meeting the following week and asks how the lesson went. Jim indicates how pleased he was with the lesson and that next year he plans to do even more with concrete materials. He also plans on extending the lesson with figures representing rhombi and squares.

The cycle continues as the supervisor checks back with the teacher. The teacher has demonstrated professional growth by indicating how the ideas can be extended to other lessons.

STANDARD 2:
TEACHERS AS PARTICIPANTS IN EVALUATION

The evaluation of the teaching of mathematics should provide ongoing opportunities for teachers to—

♦ **analyze their own teaching;**

♦ **deliberate with colleagues about their teaching;**

♦ **confer with supervisors about their teaching.**

Elaboration

The emphasis in this standard is on the teacher being a significant participant throughout the evaluation process. In particular, teachers should be given the opportunity and encouragement to engage in reflecting on and evaluating their own teaching and to discuss their teaching with colleagues who have observed their teaching. When evaluations are conducted by supervisors, it is imperative that teachers play a central role in providing information about their own teaching, including their goals and analysis of teaching. It is crucial that teachers see the evaluation process as one that contributes to their professional growth as teachers of mathematics, thereby necessitating their participation in the evaluation process.

Supervisors should establish collegial relationships with teachers in ways that foster an atmosphere for evaluation that is conducive to improving instruction. Growth is nurtured when all parties are interested in improving the teaching of mathematics and recognize that improvement is an activity for all teachers, regardless of their level of professional development. The notion of "coaching" is relevant here, since continual feedback should be furnished on progress in mastering a greater repertoire of instructional strategies. Preparation for peer coaching is an important aspect of staff development.

When a purpose of evaluation is to produce a report on a teacher's competence, teachers should be given opportunities to provide their interpretation of events and to share their instructional goals and expectations of students. Further, teachers should have access to any information accumulated during the evaluation process.

It is important that administrators create an atmosphere in which teachers are encouraged to solicit colleagues' help in guiding the improvement of instruction. Teachers should be given time to collaborate with peers and to observe and analyze each other's lessons. Even in the context of self-analysis, teachers should have confidence that the administration is supportive of change and will provide resources necessary to initiate change.

Vignettes

The teacher attends a mathematics conference as a professional development activity.

2.1 Pat Schuette, a second-grade teacher, is attending her first state mathematics conference, which is being held in a neighboring school fifty miles away. She is particularly interested in attending sessions on the teaching of geometry and spatial reasoning. Her students seem to enjoy these topics, and she enjoys teaching them. As she listens to one of the presenters, she reflects on the tasks she has given her students. She was pleased that many of the speaker's tasks using the geoboard were similar to the ones she used with her students. She is also impressed with the activities involving geometric solids, activities she rarely used with her students. As Pat listens to the rationale for helping children develop spatial perception, she realizes its importance and that she has restricted her tasks to ones involving only planar figures.

Later in the day, Pat attends a workshop for K–3 teachers on making and exploring three-dimensional figures. She welcomes the opportunity to learn more mathematics as well as obtain some practical activities she can use with her students. As the teachers explore the various activities, they engage in a lively discussion of how their students would react to the tasks and how they could use the activities in their classrooms. Many of the activities involve using D-Stix to make cubes, rectangular solids, and different kinds of pyramids. Pat enjoys making the figures and discovering Euler's formula for the relationship among the edges, vertices, and faces of the figures. She thinks her students will enjoy these activities as well.

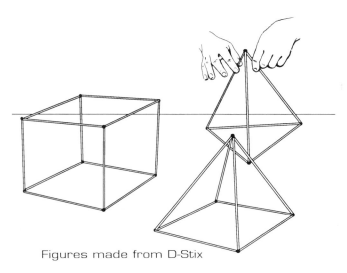

Figures made from D-Stix

Professional development consists of exchanging ideas with other teachers.

The teacher confers with her principal about obtaining materials for teaching geometry and providing students with experiences in spatial reasoning.

After the workshop, some of the teachers compare notes and share activities they feel work successfully with their students.

After she returns from the conference, Pat approaches the principal about obtaining additional materials to help teach spatial reasoning. She describes her experiences at the state mathematics conference and shows the principal the emphasis on spatial reasoning in the *Curriculum and Evaluation Standards for School Mathematics.* The principal agrees to provide a modest amount of money to purchase D-Stix and other materials that Pat and the teachers could use to promote students' understanding of geometry.

The teacher is aware of the need to incorporate calculators into his teaching but is reluctant to do so. He seeks specific suggestions from his mathematics supervisor.

2.2 Pete Wilder has been teaching eighth-grade mathematics at Blackhawk Middle School for the past ten years. Although he has read that calculators should be emphasized at the middle school level, he has been reluctant to use them. His supervisor, Tim Jackson, has been urging him to use calculators whenever possible. Tim has been disappointed that Pete has not made more progress in using calculators and in moving away from his teacher-dominated lessons. Pete admits that most of his students use calculators only for checking computations with whole numbers or decimals.

The teacher analyzes the lesson and identifies one of the problems— namely, that the calculators do not all work the same way and that he is not clear on how each of the calculators works.

The supervisor is supportive as he recognizes the need to reduce the teacher's anxiety. He also wants to help the teacher place less emphasis on computation.

The teacher and supervisor work in a collegial manner to address the problem. Assistance is obtained from another teacher.

Pete agrees to make a greater effort to use calculators if Tim will come to his class and make specific suggestions on how they can be used in a reasonable way. On the day Tim observes the class, the students are doing computational problems using the memory key on the calculator. It becomes clear that the memory functions do not all work the same way on the different calculators. This causes Pete problems in conducting the lesson. It is clear that Pete is not happy with the lesson. After the lesson he expresses his frustration to Tim and indicates that lessons like this are one of the reasons he hesitates to use calculators with his students.

Later in the day, Pete and Tim meet to discuss the problem. Pete is quite discouraged, which contributes to his anxiety in using calculators. Tim has two concerns. First, he wants to help Pete feel more comfortable and confident in using calculators. Second, he wants Pete to use calculators to explore more substantial mathematics.

Tim and Pete decide that they need more information on what his students know about using the calculator. Pete remembers that Juanita Criswell, another eighth-grade teacher, developed an activity sheet that can be used to assess students' understanding and competence in using different calculator functions. Pete is sure that Juanita would be willing to let him use the activity sheet with his students. Pete and Tim agree that this is a good idea. Tim also agrees to see if he can obtain district funds to buy calculators.

The supervisor suggests specific calculator uses that go beyond checking computations.

Tim and Pete spend the rest of the time talking about how the calculator could be used to explore mathematics rather than just use it for checking computational exercises. Pete indicates that the next unit will be on statistics—finding medians and means. Tim suggests using the calculator to investigate the following types of problems:

◆ Suppose ten students had test scores of 68, 73, 77, 81, 84, 88, 89, 91, 94, and 95. What is the average score? Suppose the teacher made a mistake and each student should receive an additional 3 points. What will the new average be? Make a guess before doing the calculation.

◆ If five workers each earn $32 500 a year and one of the workers gets a $5 000 raise, how much will the average salary increase?

♦ If four out of six workers earn $26 000, $32 000, $27 000, and $31 000 and the average for the six workers is $30 000, how much do the fifth and sixth workers each make?

Pete comments that these kinds of questions are quite different from those in the textbook. He likes them and indicates that he is willing to involve students in such explorations using calculators once he knows more about how well the students can use calculators.

The teacher indicates that the problems seem reasonable and interesting; he is willing to try to use calculators in a more substantive way.

STANDARD 3:
SOURCES OF INFORMATION

Evaluation of the teaching of mathematics should be based on information from a variety of sources including—

♦ *the teacher's goals and expectations for student learning;*

♦ *the teacher's plans for achieving these goals;*

♦ *the teacher's portfolio, consisting of a sample of lesson plans, student activities and materials, and means of assessing students' understanding of mathematics;*

♦ *analyses of multiple episodes of classroom teaching;*

♦ *the teacher's analysis of classroom teaching;*

♦ *evidence of students' understanding of, and disposition to do, mathematics.*

Elaboration

Any evaluation of teaching should be based on multiple observations and a variety of types of data—not on a single observation or a single source or type of information. A teacher's goals and expectations for student learning should be part of the information used in the evaluation process and should be discussed with the teacher prior to classroom observations. It is important that any evaluation maintain a longitudinal and cyclical orientation by considering goals previously established and those looking to the future.

Evaluation should be based on observations of the teacher teaching mathematics in a variety of contexts—that is, at different grade levels, with a variety of students, and across mathematical topics. Such variety provides a sound basis for evaluating a teacher's expertise in using various teaching methods. Some visitations should be on consecutive days to understand better the continuity of classroom events. As suggested in Standard 1, these sources of information can be used either by a teacher engaged in self-assessment or by colleagues and supervisors working collegially with the teacher.

Clearly it is unreasonable to expect that all possible teaching situations for a given teacher can be observed. Accordingly, a sample of a teacher's lesson plans, student activities and materials, projects, and student assessment techniques that the teacher has gathered over a period of time should supplement observations of the teaching process. The portfolio should consist of an ongoing collection of sample materials that provide information about "life in this teacher's classroom." The portfolio can provide a basis for self-analysis or analysis and discussion with peers or supervisors.

A teacher's own analysis of the teaching process could provide valuable information about what the teacher is intending to accomplish and could also provide a basis for improvement.

Evidence of students' understanding of, and disposition to do, mathematics should provide *a* source of information about teaching, but it should not be the only source of information. Furthermore, learning and disposition should be considered with respect to every student; that is, increasing the learning of mathematics and promoting a disposition to do mathematics for some students at the expense of neglecting other students is not appropriate. Finally, an assessment of students' learning and disposition to do mathematics should be based on the full range of mathematical activity described in the NCTM *Curriculum and Evaluation Standards for School Mathematics*; it should not be based on a narrow range of specified objectives.

Vignettes

3.1 Sandi Olson, the principal at Westside Middle School, is visiting Ben Bede's seventh-grade mathematics class. Before observing the class, she had talked with Ben about his plans for the class. She notes that he is always well organized. However, her first observation during the third week of school left her with the impression that his teaching rarely goes beyond what is presented in the textbook. At least the examples and activities he used were from the textbook. Still, she is impressed with Ben's ability to describe individual students and to diagnose individual learning problems. He seems to know his students very well—especially since it is only October.

In the conference before the observation, the principal and teacher confer about the teacher's plans.

The principal is forming an opinion on how the teacher teaches, but the source of information is limited.

Ben begins class by covering the homework on changing written statements to simple equations. After answering all the questions, Ben introduces the new material involving writing equations based on data presented in a table using the examples in the text. He has the students work a few problems and then assigns problems from the text. The balance of the period is spent on students doing the homework assignment.

Mrs. Olson makes a note for their planned after-school conference to encourage Mr. Bede to be more creative, since the lesson was very routine. On the checklist, she is contemplating giving Ben high marks for organization, knowledge of the content, and general manner in conducting the class. She observes that the room is attractively arranged with interesting bulletin boards—suggesting some degree of creativity. She is considering giving him a below-average rating on his ability to facilitate his students' mathematical disposition or to encourage problem solving or communication in the classroom. She is generally unimpressed with the development of new material.

The principal is forming conclusions based on the observation of one class.

The bell rings and students begin to pack up and leave. Mrs. Olson is ready to leave as well when suddenly there is a rush of students coming into the room for the next class. They are quite excited. Mrs. Olson decides to stay a few more minutes and see what is going on.

Maria has a branch of a poplar tree, Hector has a branch of a weeping willow tree, and Paul has a branch of an almond tree.

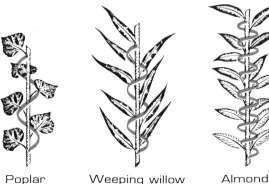

Poplar   Weeping willow   Almond

Other students have objects that represent spirals—sea shells, pine cones, and sunflower heads. Mrs. Olson hears one of the students proclaim, "Oh! Look at that one! Cool!" Mrs. Olson decides she needs to observe this second class in addition to the first class.

The principal decides that her evaluation should include the observation of a second class.

The students move their chairs into small groups as class begins, and as Mr. Bede has requested, they begin talking about the way the leaves are arranged on the branches. The students are a little confused about how any of this relates to mathematics. Mr. Bede asks them to note how the leaves are arranged on the branches. After some discussion the students observe that the numbers 3, 5, and 8 correspond to the arrangements of the leaves. Mr. Bede asks if these numbers mean anything. Hector remembers that the numbers are part of the Fibonacci sequence, which they had studied earlier. Several students ask if this is luck or whether mathematics relates to other things in life. With a twinkle in his eye, Mr. Bede asks them to take out the examples of spirals that they brought to class.

The teacher is helping students make connections between mathematics and natural phenomena.

The connection between the arrangement of the leaves and the Fibonacci sequence is noted.

The teacher will point out other examples of the Fibonacci sequence.

About this time Mrs. Olson is paged to come to the office to meet with the superintendent. She had forgotten the appointment, since she had become so fascinated with Ben's lesson. His efforts to make mathematics interesting and the students' obvious excitement has not gone unnoticed by Mrs. Olson. She is sorry that she is late for the appointment, but she is so very pleased that she stayed a few minutes to observe Ben's second-period class. What a difference it will make in her evaluation of his teaching. She is looking forward to discussing the two lessons with Mr. Bede after school and benefiting from his analysis.

The principal notes the teacher's efforts to make mathematical connections and to help students' develop a disposition to do mathematics.

The principal realizes that evaluation of teaching should be based on observing more than a single class. She will also use the teacher's analysis of the lesson.

3.2 Doug Reid teaches first-year algebra at Jefferson High School. He voices concern to his mathematics supervisor, Yolanda Hernandez, that his total evaluation last year was based on only two classroom observations. Yolanda is concerned about this also, but she explains that her responsibilities make it virtually impossible for her to visit each teacher more than two or three times a year. During the conversation, she indicates a willingness to base her evaluation on information provided by the teachers, if they can suggest a reasonable way of doing this.

A teacher expresses a concern regarding the evaluation process. His mathematics supervisor shares the concern and is willing to work collegially with the teachers to help resolve the problem.

After considerable deliberation, the teachers and Yolanda decide that the teachers will provide portfolios for her to consider and that she will base part of her evaluation on these portfolios. The teachers decide that the portfolios will contain sample lesson plans and student activity sheets they have used, various problems they have used for "warm-ups" or "attention-grabbers," special materials they have created for students, chapter quizzes and tests, notes on computer software and other materials, and a description of what they consider to be their best lessons.

The teachers and the supervisor decide to develop portfolios containing a variety of sources of information for the supervisor's consideration.

Doug is skeptical about the idea. He thinks the creation of a portfolio will be a lot of extra work that won't really help him improve as a teacher. Yolanda emphasizes that the teachers should not prepare additional materials but rather should see the portfolio as a means of collecting information on teaching that has already occurred. That is, the portfolio should be seen as a means of *collecting* information rather than *creating* information. Doug thinks this is an important point; it eases his mind that he won't have to be doing "busy work" for the sake of evaluating his teaching.

The teacher questions whether generating a portfolio will keep him from doing the more important things in teaching.

As the portfolio develops, Doug comments, "I'm beginning to see the value of the portfolio. Although it takes a little time to select and compile lesson plans, activities, and chapter tests, it turns out to be two-pronged—it is valuable for my own analysis as well as for the supervisor's. Yolanda can use this to assess my short- and long-range goals and provide feedback on my quizzes and tests. Actually, it is less work than I had originally imagined."

Yolanda indicates that the portfolios are helpful in keeping her in touch with what the teachers are doing. Actually, she is surprised that they provide as much information as they do. She notes, for example, that Doug's algebra tests tend to overemphasize computational algorithms. She offers several alternative items that he can use in subsequent testing. She likes his warm-up activities and compliments him on the way he begins the algebra classes.

At a faculty meeting later that month, the principal compliments the department for its innovative approach to improving instruction. She is particularly pleased that the teachers and the supervisor have worked together in a professional way in developing procedures for evaluation.

The teacher recognizes the value of self-analysis as part of the process of periodic evaluation.

A teacher's goals and plans should be sources of information for evaluation.

The portfolios provide information about the teaching process to which the supervisor can react and provide suggestions for improvement.

The principal is supportive of the new approach.

Summary: Process of Evaluation

The first three evaluation standards are based on the assumptions stated at the beginning of this section. The process of evaluation should reflect that the overall intent is to improve instruction, that it should be a dynamic and continual process, that teachers should be an integral part of that process, and that because of the complexity of teaching, it should involve a variety of sources of information gathered in various ways. The standards emphasize that teachers should be encouraged and supported to engage in self-analysis and to work with colleagues in improving their teaching. When evaluation involves supervisors or administrators, their relationships with teachers should be collegial with the intent to improve instruction.

The following five standards focus on what should be observed during the evaluation process. They assume that observers, including teachers themselves, should have a framework from which to observe classroom activities. These standards provide that framework. The framework is emphasized in the vision of teaching presented in the first section of this volume and is rooted in the *Curriculum and Evaluation Standards for School Mathematics.* The standards involve both mathematical content and processes.

THE FOCI OF EVALUATION

The standards in this section emphasize the type of information that should be obtained in assessing classroom teaching. At times the assessment process should focus on what the teacher is doing, and at other times it should focus on what the students are doing. By focusing on the teacher, assessment can determine the teacher's command of knowledge and strategies for teaching mathematics as well as whether the teacher is providing adequate encouragement for students to learn. By focusing on the students, assessment can determine whether the teacher has provided a context and opportunity for students to be engaged in significant and appropriate activities and whether the students demonstrate a disposition to do mathematics.

To a great extent, improving the teaching of mathematics depends both on a teacher's ability to determine what individual students know and how they construct mathematical ideas and on the teacher's ability to base instruction on those determinations. Teachers must be able to analyze their students' understanding of both mathematical content and mathematical processes. Teachers must also be able to analyze how well groups of students reason and solve problems together and communicate their mathematical ideas. Such appraisal is vital to using group work in ways that foster the development of students' mathematical power within the classroom community. The teacher's ability to analyze and describe group as well as individual learning is fundamental to the following standards.

STANDARD 4:
MATHEMATICAL CONCEPTS, PROCEDURES, AND CONNECTIONS

Assessment of the teaching of mathematical concepts, procedures, and connections should provide evidence that the teacher—

◆ **demonstrates a sound knowledge of mathematical concepts and procedures;**

◆ **represents mathematics as a network of interconnected concepts and procedures;**

◆ **emphasizes connections between mathematics and other disciplines and connections to daily living;**

◆ **engages students in tasks that promote the understanding of mathematical concepts, procedures, and connections;**

◆ **engages students in mathematical discourse that extends their understanding of mathematical concepts, procedures, and connections.**

Elaboration

The primary emphasis in this standard is on the teaching of mathematical content. The teacher should demonstrate a deep understanding of mathematical concepts and principles, connections between concepts and procedures, connections across mathematical topics (e.g., providing geometric interpretations of probability concepts or of factoring whole numbers), and connections between mathematics and other disciplines. A teacher who has a sound knowledge of mathematics can respond appropriately to students' questions, can design appropriate learning activities involving a variety of mathematical representations, and can orchestrate mathematical discourse in the classroom. Furthermore, the demonstration of a sound knowledge of mathematics includes the orientation that mathematics is, and continues to be, the result of human endeavor and that the uses of mathematics permeate modern life. On the contrary, making frequent mathematical mistakes, using limited or inappropriate representations, or presenting mathematics as a static subject whose meaning is derived solely from symbolic representations suggests that the teacher does not have an acceptable command of mathematics.

The teacher should engage students in a series of tasks that involve interrelationships among mathematical concepts and procedures. The acquisition of mathematical concepts and procedures means little if the content is learned in an isolated way in which connections among the various mathematical topics are neglected. Instruction should not be limited to a narrow range of outcomes, such as memorizing definitions or executing computational algorithms. Instead, instruction should incorporate a wide range of objectives as suggested in the *Curriculum and Evaluation Standards for School Mathematics*. Further, the teacher should emphasize mathematical communication with the intent of expanding students' understanding of mathematical content and connections.

Connections should occur frequently enough to influence students' beliefs about the value of mathematics in society and its contributions to other disciplines. Regardless of what mathematics is being studied, students should have the opportunity to apply the mathematics they have learned to real-world situations that go beyond the usual textbook word problems. Students should see mathematics as something that permeates society and, indeed, their own lives. This standard implies that instructional activities aimed at promoting students' appreciation of mathematical connections should take advantage of students' experiences and interests.

Vignettes

The teacher reflects on her previous teaching of fractions and decides to create new tasks involving multiple representations of fractions.

The teacher notes her students' performance.

4.1 Sara Rasmussen has been teaching seventh-grade mathematics for several years. Before that she taught mathematics to fifth and sixth graders. Although Sara is an excellent teacher, she is continually concerned about her teaching of common fractions, particularly with her students' ability to interpret fractions in a variety of contexts and to interpret various operations with fractions. This year she has made a special effort to create tasks that require the use of a variety of representations of common fractions, including the number line, regions, parts of sets, decimals, and measurement. She feels that the tasks have helped the students develop a good grasp of translating among representations—representing ¾ as a region within a rectangle, as a point on the number line, and as a decimal, for example. She is also pleased that her students are proficient in adding fractions.

The teacher observes that students have difficulty making the connection between the concept of fraction and adding fractions using the number line.

The teacher adjusts instruction to help students make the connection.

Sara decides to see how well the students can make the connection between the concept of fraction and the addition of fractions. She asks them to add ¾ + ½ and to provide an interpretation of finding the sum using the number line. She is surprised that the students have very little sense of how to interpret the addition of fractions using the number line. They can mark the points ¾ and ½ and the sum 1¼, but they fail to make the connection with finding the sum by starting at the point ¾ and moving ½ of a unit to the right to obtain 1¼. She spends the greater part of one period helping the students understand this connection.

The teacher discusses the problem with a colleague, who offers a suggestion.

Later in the day Sara is talking with another teacher about the problem to see if she has any suggestions for activities that could extend what she has started. Her colleague suggests a task that requires students to reason why certain procedures for adding fractions don't work. Sara decides she will try the activity.

The next day Sara writes the following examples on the board:

The teacher provides students with an opportunity to use their mathematical reasoning in a situation involving concepts and procedures.

$$\frac{1}{2} + \frac{3}{6} \overset{?}{=} \frac{4}{8} \qquad\qquad \frac{1}{3} + \frac{1}{3} \overset{?}{=} \frac{2}{6}$$

$$\frac{1}{5} + \frac{1}{2} \overset{?}{=} \frac{2}{7} \qquad\qquad \frac{0}{4} + \frac{1}{5} \overset{?}{=} \frac{1}{9}$$

She asks the students to copy the examples into their journals and to write a brief explanation on whether they think the examples are correct or not.

A few days later, Sara collects the journals and observes the following entries:

◆ It couldn't be right because ½ plus ½ must be more than ½.

- You need like terms—like 1 centimeter and 2 millimeters is not 3 centimeters. They need to be all in centimeters or all in millimeters.
- It is a good way to add fractions because it is easy.
- If you take one half a pie and one half a pie you get a whole pie, not part of a pie.
- If you start off at ⅕ and don't add anything you wouldn't go back to ⅑. You would stay put at ⅕.
- I used my calculator and used decimals. It gave me 0.7 for 0.2 + 0.5. That method can't be right.

She is impressed with the depth of some of the students' understanding. She is pleased that they have considered the examples in light of the various representations of fractions they have been studying. Most of the students used the representations appropriately to show why the procedure of adding numerators and denominators doesn't work.

The teacher analyzes the students' explanations.

The next day Sara divides the class into small groups and hands out the journal entries above after the students agree that it is okay to share them. She asks the students to indicate whether they agree or disagree with each of the statements and why they agree or disagree. The students discuss their reactions within their group and then share their reactions with the class. Sara feels that most of the students are making excellent contributions in analyzing the statements. One of the students illustrates the problem with the first example using both the number line and regions of a circle. Sara is impressed.

The teacher provides opportunities for students to discuss the connections they have made.

The teacher observes that one student uses multiple representations.

Several days later Sara listens to a student explain that you couldn't add ⅓ and ⅚ unless you found a common unit for thirds and sixths. She conjectures that these verbalizations may have been stimulated by the previous discussions on the journal entries. She feels good about this.

The teacher observes that the students may be making connections with other lessons.

4.2 Steve Cooper has taught at North High School for ten years. He generally uses his seventh-hour planning period to write his lesson plans for the next day, but today he is meeting with the mathematics supervisor, Mr. Johnson, in preparation for a scheduled observation. Mr. Cooper's concern about tomorrow's algebra class dominates the discussion. The topic is writing the equation of a line given the coordinates of two points on the line. Typically he has not been able to make the topic interesting to the students, and, perhaps as a consequence, the students find it difficult. Mr. Johnson suggests connecting the lesson to an activity involving statistics and using a computer to graph a data set and determine a line of best fit. Together they plan the lesson and discuss some questions Mr. Cooper might use during the lesson.

The teacher has identified a problem and seeks help from his mathematics supervisor.

The supervisor suggests connecting the topic to a real-world situation and offers suggestions for increasing student involvement.

The next day Mr. Cooper begins class by asking students to guess which is longer, their foot or the inside of their arm from the wrist to elbow. The students measure and record the lengths to the nearest centimeter. Mr. Johnson notes the improvement in student involvement and interest.

The teacher begins the lesson with a real-world example. By having the students guess, he increases student involvement.

Data for the entire class, including Mr. Cooper, are recorded on the chalkboard in tabular form. Mr. Cooper then asks the students to graph a scatter plot of the data with arm length on the x-axis and foot length on the y-axis. Rob and Karen enter the data into a computer at the front of the room, displaying the data and the scatter plot on the overhead screen. The other students compare their graphs to the graph on the

The supervisor observes that the students are using the data in different formats, illustrating mathematics as a network of interconnected concepts and procedures. A computerized representation facilitates discourse.

The supervisor sees the effects of their previous planning in the use of stimulating questions connecting mathematics to the real world.

The teacher connects writing the equation of a line to real data.

overhead screen. Mazie notes that points that appear more than once on the table appear only once on the screen.

Mr. Cooper asks the students whether they could make a prediction about the length of a foot of a person with a 35-centimeter-long forearm. In a spirited discussion the students agree that, in general, people with longer forearms have longer feet, but it is difficult to make a numerical estimate from these data. Tim proposes that since ten of the people have the same arm and foot lengths, 35 centimeters would be a reasonable guess. Mr. Cooper uses his pen to highlight the two points (22, 22) and (25, 25) on the projector and then lays his pencil across those two points. The students notice that most of the points are on or close to that line. After noting that the equation of this line is $y = x$ because every point has equal x and y coordinates, Mr. Cooper helps the students understand how to find the equation of the line through (23, 23) and (26, 25). He then asks the students to pick two points on their scatter plots and follow the same technique to produce the equation of their line. The students compare their equations. To conclude the lesson, Mr. Cooper has the

Algebra Class		
Student	Arm (cm)	Foot (cm)
1	29	29
2	23	20
3	24	23
4	23	23
5	26	25
6	21	23
7	23	23
8	24	24
9	24	24
10	25	25
11	27	26
12	22	21
13	27	24
14	22	22
15	25	24
16	23	23
17	23	23
18	26	25
19	24	26
20	24	24
21	22	23
22	25	26
23	29	25

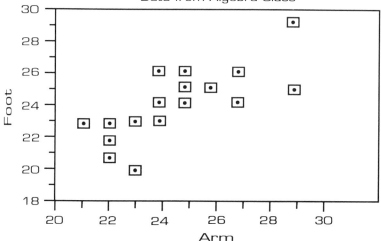

Data from Algebra Class

computer generate the equation of the "best fitting" line, $y = 0.65424x + 7.9987$, and graph the line on the screen. The bell rings before Mr. Cooper can have the students calculate the predicted length of a foot of a person with a 35-centimeter forearm using the computer-generated equation.

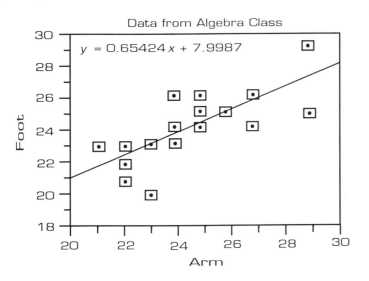

Later that day, during his seventh-hour planning period, Mr. Cooper and Mr. Johnson discuss the lesson.

Mr. Cooper: This seemed to go better than the textbook approach I have used in the past.

Mr. Johnson: The students were very engaged.

Mr. Cooper: Yes, but they will still need lots of practice to get the procedure down pat.

Mr. Johnson: Some of that practice should include situations like you used today. The students learned much more than how to write the equation of a line.

Mr. Cooper: It takes so long to do a lesson like this. We didn't get finished in a class period.

Mr. Johnson: Student interaction in inquiry lessons like this one take more time, but the students really seemed to understand what was

The supervisor and teacher confer on the lesson. The teacher notes that students have a more positive attitude toward the topic but need more practice.

happening. I think they will remember this much longer than a chalk-board explanation.

Mr. Cooper: I can probably do a better job of managing such a lesson next time.

The supervisor comments on the strengths of the lesson and offers encouragement.

Mr. Johnson: You'll find that things get easier with experience. Keep in mind the high level of student interest and the mathematical potential of the lesson. It was an excellent first lesson with this new material and approach.

STANDARD 5:
MATHEMATICS AS PROBLEM SOLVING, REASONING, AND COMMUNICATION

Assessment of teaching mathematics as a process involving problem solving, reasoning, and communication should provide evidence that the teacher—

- *models and emphasizes aspects of problem solving, including formulating and posing problems, solving problems using different strategies, verifying and interpreting results, and generalizing solutions;*

- *demonstrates and emphasizes the role of mathematical reasoning;*

- *models and emphasizes mathematical communication using written, oral, and visual forms;*

- *engages students in tasks that involve problem solving, reasoning, and communication;*

- *engages students in mathematical discourse that extends their understanding of problem solving and their capacity to reason and communicate mathematically.*

Elaboration

Problem solving, reasoning, and communication are processes that should pervade all mathematics instruction and should be modeled by teachers. Students should be engaged in mathematical tasks and discourse that require problem solving, reasoning, and communication. Consequently, assessing the teaching of mathematics should determine whether teachers and students are actively involved in these processes. The acquisition of the ability to represent mathematics in these ways takes place over time and hence should be a continuing focus of instruction. It follows that assessing the existence of these processes in the teaching of mathematics must similarly take place over time.

Teaching mathematics from a problem-solving perspective entails more than solving nonroutine but often isolated problems or typical textbook types of problems. It involves the notion that the very essence of studying mathematics is itself an exercise in exploring, conjecturing, examining, and testing—all aspects of problem solving. Tasks should be created and presented that are accessible to students and extend their knowledge of mathematics and problem solving. Students should be given opportunities to formulate problems from given situations and create new problems by modifying the conditions of a given problem.

Teachers should engage students in mathematical discourse about problem solving. This includes discussing different solutions and solution strategies for a given problem, how solutions can be extended and generalized, and different kinds of problems that can be created from a given situation. All students should be made to feel that they have something to contribute to the discussion of a problem. Assessment should focus on the notion of whether mathematics is being taught in such a way as to promote these aspects of problem solving.

Teaching mathematics as an exercise in reasoning should also be commonplace in the classroom. Students should have frequent opportunities to engage in mathematical discussions in which reasoning is valued. Students should be encouraged to explain their reasoning process for reaching a given conclusion or to justify why their particular approach to a problem is appropriate. The goal of emphasizing reasoning in the teaching of mathematics is to empower students to reach conclusions and justify statements on their own rather than to rely solely on the authority of a teacher or textbook.

Assessment should seek evidence that students are using inductive reasoning, proportional reasoning, and spatial reasoning and are constructing arguments. Assessing whether mathematics is being represented as a process of reasoning should focus on whether the teacher demonstrates the pervasiveness of mathematical reasoning throughout all areas of mathematics and whether the teacher requires students to use various reasoning processes.

Communication is the vehicle by which teachers and students can appreciate mathematics as the processes of problem solving and reasoning. But communication is also important in itself, since students must learn to describe phenomena through various written, oral, and visual forms. The notion of communication emphasized in this standard cannot be fully realized in a lecture-oriented lesson or when students' responses are limited to short answers to lower-order questions. This standard suggests that mathematics is learned in a social context, one in which discussing ideas is valued. Classrooms should be characterized by conversations about mathematics among students and between students and the teacher.

Mathematical communication can occur when students work in cooperative groups, when a student explains an algorithm for solving equations, when a student presents a unique method for solving a problem, when a student constructs and explains a graphical representation of real-world phenomena, or when a student offers a conjecture about geometric figures. A teacher should monitor students' use of mathematical language to help develop their ability to communicate mathematics. This could be done by asking students if they agree with another student's explanation or by having students provide various representations of mathematical ideas or real-world phenomena. The emphasis should be on all students communicating mathematics, not just on the more vocal students. In order for teachers to maximize communication with and among students, they should minimize the amount of time they themselves dominate classroom discussions.

Vignettes

The teacher constructs her lesson on the basis of her students' previous experiences. In selecting the task, she considers its potential for fostering mathematical reasoning and communication.

5.1 Pat Kowalczyk's kindergarten class enjoys activities involving continuing patterns that have been started using blocks, beads, themselves, and other items. Today Mrs. K, as the children call her, plans on having her class construct patterns using their names. She thinks that this activity will extend the work she has been doing to encourage them to reason and communicate about mathematics with one another. She has prepared a paper with a 5 x 5 grid of 2-centimeter squares for each student.

At their tables the students fill out the grid, using one square for each letter of their name. When they finish writing their names the first time, they start over and continue until each of the 25 squares contains a letter.

K	E	N	T	K
E	N	T	K	E
N	T	K	E	N
T	K	E	N	T
K	E	N	T	K

Mrs K: Select your favorite crayon and color in the squares that contain the first letter of your name.

Mrs. K walks around the room observing and listening to the students as they work. When Susan wants to know if she should color both the S's in her name, Mrs. K responds with a question, "Are both the first letter of your name?" Susan thinks for a moment and then says, "No, only this one is," and she colors only the first S in Susan. Mrs. K makes a mental note that Susan seems confident in her decisions and does not seek additional confirmation from her. As she continues to walk around, Mrs. K observes that some children seem to understand the activity and work independently, some are actively conferring with others, and some are waiting for her to help them. She muses, not for the first time, about what more she could be doing to foster greater self-reliance by her students.

In assessing her teaching, the teacher focuses on the students' ability to rely on their own mathematical reasoning.

When the students complete their grids, Mrs. K asks the class if they can predict who has the same patterns of colored-in squares on their grids. She tries to phrase the question so as to encourage the students to reason and to communicate their ideas. She notices that she is improving in her ability to construct good questions on the spot.

The teacher poses questions that engage students in mathematical reasoning and communication. She then analyzes her own ability to ask questions.

The students quickly guess that the two Jennifers in the class should have the same pattern. Mrs. K asks several students to explain how they can be sure of this without even checking the girls' grids. When she hears Marcus say, "'Cause they have the same name so their papers gotta be the same too," she is really pleased. Calling on him more often seems to be paying off.

Searching for the next good question, Mrs. K challenges the students to find similar patterns where the students do not have the same first name. After some checking around, the students find that Kent's and Kyle's grids have the same pattern.

Kent: Maybe names that begin with the same letter look the same.

Mrs. K: Is there anyone else whose name begins with the letter *K*? (Katrina, Kathy, and Kevin all jump up, waving their hands.)

Katrina: But my grid is different from Kent's and Kyle's.

Kathy: But mine is the same as Kevin's.

Mrs. K: Does this fit the rule that the names that begin with the same letter give the same pattern?

Students (in unison): No!

Mrs. K looks around, trying to decide on whom to call and tries to remember who has not spoken much today. She remembers that Nikki has not said anything today, although she did complete her grid quickly.

Mrs. K: Nikki, how can we change our rule so that it works?

Nikki: Well, I think it will work if they have the same number of letters and if their name begins with the same letter.

Laura (excitedly): Mine matches Kathy's, but our first letters are different.

Mrs. K: Let's check it out. (She holds them up to the window, one on top of the other.) Hey, it looks like they *do* match!

At this point, Dave, Jane, and José put their patterns by Kyle's and Kent's and are surprised that the patterns match. They don't know how to express their finding. Mrs. K is a little surprised that this is hard to explain. Judy says that it has something to do with the length of the name. Short names seem to match short names but not long names. Finally, Stanley says that the names with the same number of letters will match. Some of the other students question whether he is right. After examining many other examples, they conclude that he is correct.

After school, Mrs. K reflects on the lesson. She writes a few notes in her journal—about Marcus, Nikki, and several other students. She also writes down the task so that she can remember it for the future and indicates that she thinks it could be used profitably again. She is impressed with the students' ability to reason. She thinks that letting the students use different-colored crayons to color in the grids may have distracted them from the lesson's primary objective. She makes a note to let students pick only one color next time she uses this activity. Although she thinks she is getting better at formulating good questions, she also thinks that she needs to find more ways to encourage students to communicate their ideas with one another and to build on one another's reasoning.

5.2 The principal has given Doug Walker, an experienced seventh-grade teacher, release time to observe Louise Knight's seventh-grade mathematics class and to help her develop professionally. Ms. Knight is a young teacher who demonstrates a considerable amount of energy and creativity in her teaching. The day before the scheduled visit, Mr. Walker talks with Ms. Knight about the upcoming lesson, her objectives, activities, and her expectations for the students. The focus of the lesson is on the use of geometric terms to describe where objects are located. Ms. Knight provides Mr. Walker with a list of terms that the students are to use in writing the directions. This list includes *perpendicular bisector, angle bisector, midpoint, right angle, acute angle,* and *obtuse angle.*

The principal encourages collaboration as a professional development activity.

The teacher's goals for the lesson emphasize communication through written and visual forms.

The next day, Ms. Knight starts the lesson by reviewing the geometry terms. She then organizes the students into pairs. She carefully selects the pairs so that each student whose first language is not English will be working with a student whose first language is English. Out on the school grounds each pair determines where their treasure is hidden. They write directions for finding the treasure, using the geometry terms they had reviewed earlier. They check the accuracy of their directions by following them from the beginning to see if they arrive at the spot they had identified for the treasure. Then they draw a map that indicates the location of the treasure.

The teacher demonstrates sensitivity to students whose native language is not English.

The teacher has provided an opportunity for written, oral, and visual communication in the context of a problem-solving activity.

Mr. Walker notes how Ms. Knight checks the progress of each pair of students, helping them when necessary but not providing students with correct descriptions. She also records the specific difficulties of individual students.

The colleague observes that the teacher encourages communication but avoids placing herself as the center of attention.

The following set of directions is written by Dave and Julio:

The students write directions, a form of communication.

1. Start at the intersection of the sidewalks. Bisect the right angle and proceed 25 yards along the bisector. Stop.
2. Now go to the midpoint of the line segment between where you are and the flag pole.
3. Turn left 90 degrees and go 10 yards. Stop.
4. Go halfway to the cedar tree. Find the treasure.

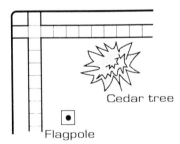

Mr. Walker accompanies Dave and Julio as they check the accuracy of their directions.

Mr. Walker is impressed with the quality of the students' mathematical communication and with the way Ms. Knight has organized the pairs of students. Most of the students used all the terms identified by Ms. Knight. He notices that several students double-check the meaning of some of the terms in their textbook. He observes that Ms. Knight spends a considerable amount of time encouraging the students to be as precise as possible in writing their directions. She usually does this by asking questions rather than by telling the students what the directions should be.

The colleague observes students engaged in mathematical discourse—discussing terms and checking references.

The colleague offers a suggestion to improve the activity.

After the lesson Ms. Knight and Mr. Walker talk about the ability of the students to write directions. They both think that although the students' directions were generally clear and reasonably concise, a few were confusing and did not lead to the treasure. Mr. Walker suggests that the students should write the directions first and then hide the treasure accordingly. This might avoid the problem of locating a point and then trying to write the directions that guides a person to that point. Ms. Knight likes this suggestion. She agrees that locating the point first and then writing the directions may have led to some confusion.

The colleague offers additional suggestions for improving the lesson.

The colleague is supportive of the teacher's creative efforts.

Mr. Walker also suggests that when a treasure is not found because of faulty directions, the "search party" describe their path to the students who wrote the directions so that the discrepancy can be discussed. Ms. Knight sees this as an added opportunity for students to communicate mathematics. She likes having the students validate the directions rather than just relying on the teacher. Mr. Walker compliments Ms. Knight on the lesson and her courage for trying an activity-oriented lesson. He wishes her well on the next day's treasure hunt.

The teacher reads professional journals to improve his teaching.

The teacher and colleague work together.

The colleague offers specific suggestions.

5.3 Art Heyen has been reading various articles in the *Mathematics Teacher* about the importance of emphasizing mathematical processes when teaching mathematics. He decides to make a concerted effort this year to incorporate these ideas into his teaching. At the beginning of the year, he meets with Diane Rowan, an experienced mathematics teacher, to discuss how his teaching could become more process oriented. Diane suggests that he start with a few selected topics to "get the feel of it" and then work from there. Diane offers suggestions for a lesson on graphing parabolas that Art could use later in the year. The emphasis is on considering the equation

$$y = ax^2 + bx + c$$

and examining the effect on the graph when different values of a, b, and c are used.

The teacher initiates contact with a colleague to observe his teaching.

The teacher notices his students' mathematical disposition has improved, perhaps because of his more process-oriented approach.

In the past, the teacher used lecture/listening as the primary instructional technique.

In November, Art is ready to teach the lesson on graphing parabolas. He invites Diane to observe the lesson and make suggestions. He indicates that he has had moderate success with other lessons in which mathematical processes have been emphasized. He complains somewhat about it taking so long to find good materials but notices that the students seem more interested in mathematics this year than any of the previous two years that he has been teaching.

Art typically teaches the lesson on graphing parabolas by modeling several graphs and helping the students locate the vertex and several other points, which are then plotted. After several demonstrations, he assigns practice problems. This year he will teach the topic with a greater emphasis on conceptual development.

The teacher incorporates technology into the lesson. He then begins by determining and building on what the students know.

Art begins the lesson by passing out graphing calculators that have just recently been obtained. He asks the students to write down three statements or words they associate with the equation

$$y = ax^2 + bx + c.$$

The colleague considers how the teacher could have improved a student's reasoning ability.

Some of the phrases are *quadratic equation*, *parabola*, and *horseshoe shaped*. One student mentions that if $x = 0$, then $y = c$. Art asks what this means, but the student is not sure. Diane thinks Art might have spent more time helping the student reason through his conclusion.

Art asks the students to use the graphing calculators to graph the cases when $a = 0.25, -0.5, 1, 4$, and -8 and b and c are both 0, and then sketch and label the graphs on the same set of axes on the graph paper he has provided.

The teacher involves students in mathematical exploration by having them consider special cases using graphing calculators.

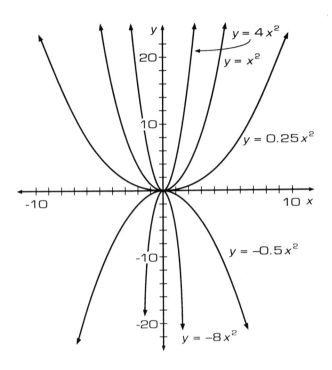

The students use these sketches to answer the following questions on the worksheet that Art has passed out.

The teacher asks the students to conjecture about the effects on graphs when changes in the value of a occur.

Use your sketches on sheet 1 to answer the following questions:
a. What property is common to all the graphs?
b. Under what condition does the graph of $y = ax^2$ open upward? Downward?
c. As the absolute value of a increases, what happens to the graph of $y = ax^2$?

The teacher's questions require the students to engage in written communication.

As the students work on the questions, Art and Diane walk around the room and check the students' progress. Diane notices that one student has drawn a horizontal line for one of the sketches. She asks if it is possible for the graph of a quadratic function to be a horizontal line. The student seems puzzled. After rechecking his figures, the student finds out that he had entered $y = 4^2$ on the calculator rather than $y = 4x^2$. A student who was listening claims that you could get a horizontal line. Diane asks her how this could happen. The student argues that it can happen when the a and b coefficients are zero because then you have a constant function. Both Art and Diane are impressed with the student's reasoning, even though they recognize an error in her thinking.

The colleague encourages students to use mathematical reasoning.

The teacher makes the decision not to pursue the question of whether the graph of a quadratic function can be a straight line. He intends to discuss this later with the students.

Art next asks the students to consider various graphs of equations having the general form $y = ax^2 + c$. They are to first use their graphing calculators and then sketch the graphs on a set of axes for the following equations:

$$y = 3x^2 + 10 \qquad y = 3x^2 - 5$$
$$y = 3x^2 + 5 \qquad y = 3x^2 - 10$$

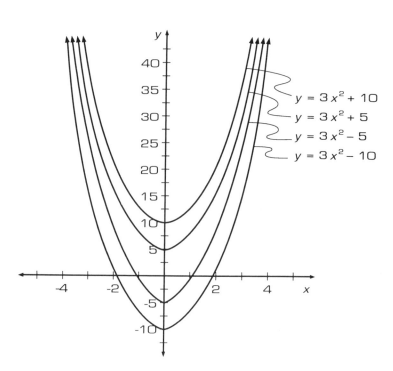

The teacher has the students use graphing calculators as tools in mathematical discourse.

The teacher asks the students to use inductive reasoning to determine the nature of the graphs.

When the sketches are completed, the students are instructed to address the following questions:

a. How does the value of c affect the graph of $y = 3x^2 + c$?

b. At what point does the graph of $y = 3x^2 - 4$ intersect the y-axis?

The teacher gives the students an opportunity to apply their generalizations in sketching other equations.

On the basis of the graphs they have sketched, the students are then asked to consider the following questions:

Without using the calculator, indicate whether the graph of each of the following equations will open upward or downward; whether the graph will be relatively narrow or wide; and where the graph intersects the y-axis. Then sketch the graph on a sheet of graph paper.

$$y = \frac{-1}{4}x^2 - 4 \qquad y = 5x^2 + 4$$

$$y = -x^2 + 7 \qquad y = \frac{1}{3}x^2 - 6$$

The teacher provides students with a different means of communicating mathematics.

To conclude the lesson, Art asks the students to write one or more statements about what they have learned about the graphs of quadratic equations. He collects these papers and assigns additional equations for the students to explore and sketch.

The teacher and the colleague reflect on the lesson.

After school, Art meets briefly with Diane to discuss the lesson. He mentions that he wanted to cover more material—particularly the relationship between the value of the discriminant and the number of x-intercepts. He realizes that it would be easier to just tell the students what he wants them to know, but he was very pleased with their ability to communicate mathematics in their written statements and to use inductive reasoning to figure out what the sketches of the last set of graphs will look like.

The colleague suggests how the teacher can enhance his students' ability to solve problems using inductive reasoning.

Diane concurs. She was particularly pleased with Art's repeated efforts to encourage the students to write statements about what they had discovered. She suggests that next time he might begin the lesson by asking the students what the graph of the equation $y = 1.5x^2 - 3x + 4.2$

(that is, an equation that the students wouldn't normally encounter) would look like. After students' conjectures are recorded, the lesson could be developed, following which the equation could be revisited to determine its graph. The students could use their knowledge developed in the lesson to determine its graph.

Art likes the suggestion. He sees this fitting in with his intention of helping students to reason mathematically. Diane thanks Art for inviting her into his classroom. She compliments him on his efforts to improve his students' ability to use mathematical processes.

The colleague is supportive.

STANDARD 6:
PROMOTING MATHEMATICAL DISPOSITION

Assessment of a teacher's fostering of students' mathematical dispositions should provide evidence that the teacher—

◆ *models a disposition to do mathematics;*

◆ *demonstrates the value of mathematics as a way of thinking and its application in other disciplines and in society;*

◆ *promotes students' confidence, flexibility, perseverance, curiosity, and inventiveness in doing mathematics through the use of appropriate tasks and by engaging students in mathematical discourse.*

Elaboration

If students are to develop a disposition to do mathematics, it is essential that the teacher communicate a love of mathematics and a spirit of doing mathematics that captures the notion that mathematics is an invention of the human mind. Sometimes this entails an exploration of a student's query or a consideration of multiple ways of solving a problem. Certainly, it involves a sense of communicating mathematical ideas. There is little value in telling students how exciting mathematics is if they are not actively engaged in doing mathematics themselves.

Using mathematics to explore real-world phenomena is one means of developing mathematical disposition. For example, students could consider sampling problems and forms of statistical inference using proportional reasoning as a means of understanding how mathematics relates to their lives. The notion of connections is central to this means of promoting mathematical disposition. Students could explore Euclidean properties on a sphere, such as the sum of the measures of the angles of a triangle, to consider the generality of those properties as another means of developing mathematical disposition.

Assessing the teacher's fostering of students' mathematical disposition should focus on whether the teacher facilitates students' flexibility, inventiveness, and perseverance in engaging mathematical tasks and on whether students demonstrate confidence in doing mathematics. Verbal cues that encourage students during instruction and supportive written comments on homework and test papers are obvious means of promoting a disposition to do mathematics. Teachers should be nonjudgmental when students give answers or present solutions to problems; teachers should help students to correct their mistakes, but mistakes should be recognized as a natural part of the learning process. Students should be expected to raise questions and challenge ideas generated by other students as well as by the teacher. Above all, students should have ample time to be active participants in doing mathematics.

The teacher is concerned with her students' dispositions to do mathematics and that they have a narrow perspective on what constitutes mathematics. The teacher is engaged in self-analysis of her teaching.

Vignettes

6.1 Stephanie Douglas, a fourth-grade teacher, feels that her students are not having enough experiences relating mathematics to other subjects or other aspects of their lives. She sees mathematics as a subject that involves many aspects of living and would like her students to share that view as well. In particular, she is concerned that her students think mathematics is basically computational in nature and not very interesting. She would like to change this.

As part of a school project on the environment, Ms. Douglas's class decides to plant a maple tree in the school yard on Arbor Day. The cost of the tree including planting is $50. The class discusses how they can raise the money for doing this. After considering several options, the class decides to collect aluminum cans—from home and ones that they find lying around in various places. The students feel good about this idea because they can make a contribution by cleaning up litter and by planting a tree for the school. Ms. Douglas compliments the students on their idea for raising money.

The teacher provides an opportunity for students to participate in solving a real-life problem.

The teacher encourages the students by building their confidence.

Several of the students check on how much can be earned by collecting aluminum cans. They find out that they can earn 40 cents a pound and that it takes about 30 cans to make a pound. The students ask how long it will take to earn the money. Ms. Douglas sees this as an excellent opportunity to become involved in the mathematics of prediction and estimation. The class determines that it will take 125 pounds of cans, or about 3750 cans, to earn the $50. Hank, one of the students, interjects that he can't drink that much soda pop. The other students think they can't get that many cans.

The teacher engages the students in mathematical discourse.

Ms. Douglas suggests that students from other classes could bring in cans and that their class would be in charge of collecting them and taking them to the recycling company. Eleanor suggests that they put a big container in the hallway where students from other classes could drop off their cans. The other students concur with Eleanor's suggestion. Ms. Douglas says that she will check with the school principal but she thinks it is a workable idea.

The teacher encourages students to propose solutions to a logistical problem.

Ms. Douglas suggests that the students make a graph to determine how long it will take and how their collection is proceeding. Since they have three months before they want to plant the tree, the students decide to consider six points on the graph—one for about every two weeks. The students plot their data points and eventually produce a graph. They decide that they need an additional graph, one that relates the number of cans to the amount of money they can raise. The students construct the following graphs:

The teacher helps students model the situation using mathematics.

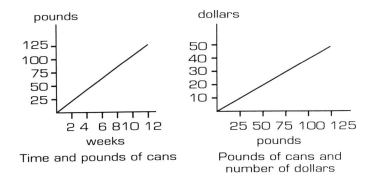

Time and pounds of cans

Pounds of cans and number of dollars

The students decide that every two weeks they will mark the graph to indicate how many pounds they have collected. After several months they construct the following graph to represent the collection process:

The teacher provides an opportunity for students to use a graph to represent real data.

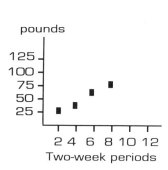

pounds

Two-week periods

The teacher reflects on the activity. She concludes that she has helped the students develop a more positive disposition to do mathematics and that the students see mathematics in a broader context than just textbook problems.

Ms. Douglas is very pleased with the activity. She thinks the students are beginning to see mathematics as more than just paper-and-pencil algorithms and that it has helped the students better understand the project. The students look forward to plotting points on the graph every two weeks. Some of the students find out that in a town in another state a store gives 45 cents a pound. They draw a graph to represent that situation and determine how much more quickly they could earn the money had they lived in that town.

The supervisor holds a preobservation conference as a first step in determining the teacher's goals for the lesson.

6.2 Mr. White is in a preobservation conference with his mathematics supervisor, Mrs. James. Mrs. James will be observing Mr. White's seventh-hour advanced algebra class tomorrow and is gathering information regarding the class. She is using the district's preobservation conference report form.

Teacher _____ Date _____

Grade/Subject _____ Observer _____

1. Learner Objectives

   a. Content (What will students be learning?)
   b. Process (How will students be learning?)
   c. Rationale (Why are students learning this content?)

2. Assessment

   a. What processes will be used to check for student understanding in class?
   b. What processes will be used to check for student understanding at the end of the lesson/unit?

3. Instructional Strategies

   a. What special resources, questioning techniques, or motivational techniques will be used?

4. Observer Focus

   a. What is the major focus of data collection?

The teacher is concerned about the mathematical disposition of the students.

In discussing his goals for the lesson on the tangent function, one of the topics in the chapter on right-triangle trigonometry, Mr. White emphasizes that he would like his students to be more confident in doing

mathematics than they are now. He indicates that he is working hard to get the students more actively involved during the lessons. Mrs. James reviews the form to make sure she is clear about what his expectations are for the lesson. She asks if there is anything in particular on which Mr. White would like her to focus during the observation. He indicates that he would like her to consider how effective he is in getting all the students involved.

The supervisor and teacher work together to make the coming observation as productive and helpful as possible.

The next day Mrs. James observes the class, and she and Mr. White sit down after school to discuss the lesson. Mrs. James begins the postobservation conference by complimenting Mr. White on his ability to get all the students involved. She could tell that he had made an extra effort to involve all the students. She then provides a factual description of his teaching: It consists primarily of small steps in which Mr. White models a concept or procedure and then engages students in practicing other similar problems. She notes that this occurred five times during the lesson. In each instance the students used Mr. White's sixty-second rule, namely, that they work on a problem for sixty seconds by themselves and then share their solutions and questions with their neighbor. She reported several instances in which students had generated their own solution methods—ones different from what Mr. White had apparently anticipated. She shared with him the following exchange at one point during the lesson.

The supervisor points out a positive aspect of the teacher's performance. She then gives a factual description of his teaching but avoids using rating terms such as good, excellent, adequate, *or* inadequate.

Mr. W: Class, find the value of x in the figure on the board.

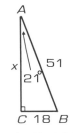

(Mrs. James reports at this point that Mr. White circled the room,

The teacher monitors students' work.

briefly stopping at desks as students worked on the problem. After two minutes, the following dialogue continued.)

Mr. W: How can we find x using the tangent ratio? Cindy?

Cindy: Well, (pause) I used the Pythagorean theorem to find it.

Mr. W: That's usually okay, but today we are focusing on the tangent ratio. How can we find x using the tangent ratio? Marc?

The students present different solution methods.

Marc: Let cos 21° = $x/51$. Then x is approximately 48—a little less, actually.

Mr. W: But, Marc, you used the cosine. How can we solve the problem using the tangent ratio? Allison?

Allison (excitedly): Well, I don't know. But I think I got a good way to solve it. Take the complement of 21 degrees and get 69 degrees; then you have sin 69° = $x/51$. You get about the same answer as Marc said.

The teacher is not supportive of students' different solution methods.

Mr. W (somewhat angry): Did anyone solve the problem using the tangent method? That's what we are studying!

Jim: Sure. Tan 21° = $18/x$.

Cindy: But isn't the Pythagorean theorem easier?

Mr. W: It probably is for this problem. But I want you to know how to use the tangent ratio.

The supervisor asks the teacher for his analysis of the data she has collected.

After reviewing this exchange, Mrs. James asks Mr. White if he sees any discrepancy between this exchange and his goals for building students' confidence in doing mathematics. In particular, she asks Mr. White to consider the case of Allison, who was obviously excited that she had discovered a way of doing the problem. Allison is not a strong student and is often a reluctant participant in class.

The teacher realizes that his actions do not help students develop a disposition to do mathematics.

At first Mr. White is defensive, but then he realizes that his actions were counterproductive to his intent of helping students develop confidence in doing mathematics. He hadn't realized that he had neglected to compliment the students for their discoveries. He wondered what he could have done differently. He confessed that he was concerned about several of the students acting up, and therefore he wasn't listening very carefully to what the other students were saying. Mrs. James suggested that they consider some alternative strategies.

The teacher reflects on the lesson and considers how he might have taught the lesson differently.

Mr. White suggests that either he should have been more accepting of the students' alternative solution methods or he should have changed the problem so that the students would have to use the tangent ratio. For example, he could have deleted the fact that side *AB* is 51 units.

The teacher recalls that he wanted to improve his students' disposition to do mathematics and that his suggestion might not help realize this objective.

With this revised problem, the students would have had to use the equation tan 21° = $18/x$. Mrs. James agrees that this would be one way to force students to use the tangent ratio. But then she asks, "What did you consider to be one of your primary objectives?" Mr. White recalls that he wanted to increase the students' confidence in doing mathematics. He agrees that narrowing the task would not necessarily contribute toward realizing this objective. Mrs. James suggests that he could change the question as follows.

Class, we want to find x in $\triangle ABC$.

How many different ways can we find a solution, including the new one we discussed today?

The supervisor provides a specific suggestion that better enables the teacher to reach one of his objectives.

Mr. White and Mrs. James discuss how this type of question can help him achieve his objective of teaching the tangent ratio but also go a long way toward promoting the students' confidence, flexibility, and inventiveness in doing mathematics. The question also emphasizes that problems sometimes have multiple ways of being solved and that students can take pride in discovering a strategy for solving a problem that nobody else in class had thought about. Allison may have been a case in point.

The importance of considering multiple ways of solving a problem is recognized as a means of promoting mathematical disposition.

Before leaving the postobservation conference, Mrs. James makes it clear that Mr. White is making significant progress as a second-year teacher. His lessons are organized, and he is genuinely interested in his students. Before Mrs. James's next visit, Mr. White agrees to work harder to create better questions and to be more sensitive to recognizing students' solutions, especially when those solutions are not the ones he was anticipating.

The supervisor is supportive of the young teacher.

The teacher reflects on his teaching and recognizes areas of needed improvement.

STANDARD 7:
ASSESSING STUDENTS' UNDERSTANDING OF MATHEMATICS

Assessing the means by which a teacher assesses students' understanding of mathematics should provide evidence that the teacher—

♦ *uses a variety of assessment methods to determine students' understanding of mathematics;*

♦ *matches assessment methods with the developmental level, the mathematical maturity, and the cultural background of the student;*

♦ *aligns assessment methods with what is taught and how it is taught;*

♦ *analyzes individual students' understanding of, and disposition to do, mathematics so that information about their mathematical development can be provided to the students, their parents, and pertinent school personnel;*

♦ *bases instruction on information obtained from assessing students' understanding of, and disposition to do, mathematics.*

Elaboration

The process of assessing teaching should determine whether and how the teacher uses evidence of students' understanding of, and disposition to do, mathematics in making instructional decisions. The assessment of students' understanding of mathematics should include methods used on a daily basis as well as those used on a less frequent basis. These methods include evaluating journals, notebooks, essays, and oral reports; evaluating students' homework, quizzes, and test papers; evaluating classroom discussions, including attention to students' mathematical problem-solving, communication, and reasoning processes; and evaluating group work, clinical interviews, and performance testing administered individually or in small groups. Such a variety of student assessment techniques reflects a sensitivity to the developmental level, maturity, and cultural diversity of the students and should provide a sound basis for creating mathematical tasks and directing mathematical discourse.

The student assessment standards of the *Curriculum and Evaluation Standards for School Mathematics* provide a basis for designing tasks to assess student understanding. Student assessment methods should also be aligned with instruction. For example, if calculators are used throughout the instructional program, then they should be allowed in testing situations as well.

Instructional activities should be based on information obtained from assessing students' mathematical understanding and disposition to do mathematics. A teacher ought to be able to determine from an analysis of evidence, for example, why a student cannot use a particular algorithm with a reasonable degree of proficiency. Does the student lack a conceptual basis for the algorithm? Is the student confused about the sequence of steps to be followed? Does the student have a sense of when to apply the algorithm, or is the algorithm applied in inappropriate contexts?

The oft-used phrase, "Are there any questions?" cannot be reliably used to determine whether students understand. However, if a student claims that a parallelogram is a quadrilateral with two sides parallel and two sides congruent and the teacher asks other students to produce a counterexample if possible, then there is an indication that the teacher is engaging students in tasks to check understanding. Another assessment technique consists of asking students to react to and evaluate another student's process for solving a problem.

As a result of student assessment, a teacher should be familiar with a student's confidence level, willingness to persevere, and other characteristics of disposition noted in the previous standard. A teacher should be able to describe individual students' mathematical dispositions beyond the general descriptions of whether a student is motivated.

Vignettes

7.1 The elementary principal, Barbara Moore, has been very impressed with the way the second-year teacher Ed Dudley conducts his first-grade class and, in particular, with the way that he makes extensive use of manipulatives when teaching mathematics.

She notices, however, that when he evaluates students' progress, he relies on paper-and-pencil tests that appear to emphasize computational outcomes. When she talks with Mr. Dudley about this, he indicates that it just seemed like a reasonable way to evaluate students—a very efficient method. He indicates, however, that he is willing to try different methods of assessing students' understanding of mathematics. Ms. Moore offers several suggestions.

The principal notices that assessment methods are not aligned with instruction.

The teacher is willing to consider a variety of assessment methods.

A week later, Ms. Moore drops by Mr. Dudley's class to see if he has had an opportunity to try any of the suggestions. She is pleased to note that in one part of the classroom Mr. Dudley is interviewing students while in another part of the room pairs of students are playing a numeration game with cubes that can be linked together and a spinner that determines the number of cubes each student receives. The students are to link the cubes together whenever they have a group of ten cubes. The students take turns spinning until each has had five turns. After the five spins they are to write the number of cubes down on a sheet of paper. Each student checks to see whether the other student has written down the correct number of cubes. Ms. Moore is quite impressed with the game. She plans to suggest to Mr. Dudley that the activity could provide him with an excellent means of assessing students' understanding of place value.

The teacher is interviewing students to assess their mathematical understanding.

The principal notes a technique that could be used for assessment purposes.

Ms. Moore decides to observe one of the interviews that Mr. Dudley is conducting. She observes the following exchanges:

Mr. Dudley gives two students, Jo and Annette, a different number of counting sticks. Each student is to bundle her sticks into groups of ten.

Jo: I have 3 tens and 4 ones.

Mr. D: Is that the same as thirty-four? (Jo hesitates. Mr. Dudley observes that she seems unsure whether the representation of 3 tens and 4 ones also represents thirty-four.)

Annette: I think they are the same.

Jo: I don't know. Let me count them. (Jo unbundles the sticks and begins counting by ones.)

The teacher assesses the student's understanding of place value and notes a possible difficulty.

Annette: I have 4 tens and 2 ones. That's forty-two.

Mr. D: How do you know?

Annette: Look. (Points to bundles of ten.) Ten, twenty, thirty, forty, now (pointing to single sticks) forty-one, forty-two. See!

Mr. D: That's very good, Annette. Let's see if we can help Jo. Jo, how are you coming?

Jo: I counted and got thirty-four. They must be the same.

After Mr. Dudley finishes his interviews, he discusses the class with Ms. Moore. She compliments him on assessing students' understanding in much the same way he teaches mathematics. He indicates that the interviews did take some extra time but that generally they took less time than he had imagined. He is very pleased with how much he learned about each student's thinking about place value during the interviews. Ms. Moore helps Mr. Dudley develop a chart to make his assessment more systematic.

Place Value and Counting

Jo	Accurately counted by ones and bundled them into tens. Not sure what number was represented, however. Needs more work on recognizing number when given representation. [Jan. 21]
Annette	Appears confident. Has good command of translating between written number and representation using sticks. [Jan. 21]

Ms. Moore compliments Mr. Dudley on his willingness to try something new and how he organized the class so that all the students were involved in learning activities. She points out that the game could also serve as an excellent vehicle for assessing students' understanding. Mr. Dudley indicates that he wants to provide Jo with more opportunities to count objects and to do activities to develop her understanding of place value.

7.2 In the mathematics department at West High School "teacher teams" have the responsibility for the development and monitoring of program and curricular changes. The "algebra team" of Art Washington, John Nystrom, and Katie Cusciaro teach all ten of the first-year algebra classes at West. Through a paid summer curriculum project, their principal, Simone Richardson, has asked them to develop a plan to reduce the high number of students who discontinue taking mathematics after first-year algebra. The teachers are concerned as well, since they have noted the high number of D's and F's that students have been receiving in algebra. Ms. Richardson asks the team to pay particular attention to the diverse backgrounds of the students at West.

In July the three teachers meet to discuss the problem. They agree that there are many factors that contribute to poor student performance and their high rate of dropping out of mathematics. They decide to focus on their teaching and assessing techniques as a first step to improving the situation. With respect to assessment, the teachers share their tests, quizzes, and the means by which they assign grades. Art indicates that many students do not complete his tests—they skip many items. He suspects that since English is not the first language for many of the

The teacher notes that the student has a good grasp of place value.

The principal compliments the teacher on aligning his assessment with his instruction.

The principal suggests a chart to help the teacher organize his assessment of individual students' progress.

The principal supports the teacher's efforts to try new techniques.

The teacher indicates he will base his instruction on what he learned during the interview process.

The teachers are empowered with the authority and responsibility to shape programs and curricula.

The teachers recognize that problems related to assessment and grading may be contributing to a broader problem.

The school district provides funds during the summer for teachers to meet and address problems.

students, they have difficulty reading some of the questions. He says that about 80 percent of his grades are based on tests and quizzes and usually the students do not do well on tests and quizzes. Yet, he feels that they demonstrate a reasonable understanding during class discussions.

The teacher realizes an inconsistency between his informal assessments during class discussions and the more formal assessments using tests and quizzes.

Katie is frustrated as well. She tries to create items in which the students are required to "explain" or "draw and label." She shares the following items.

1. Draw and label altitude $\overline{AG}$ for $\triangle ABC$. Explain why it is an altitude.

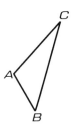

2. A student claims that x^2 is always larger than x. Is she correct? Explain your reasoning.

3. Draw and label a rectangle whose area is $(x^2 + x)$ cm^2.

She states that her students have a great deal of difficulty in writing mathematics—perhaps because of the language problem. Katie indicates that she has allowed students to work together when solving problems but not when taking tests or quizzes.

The teacher emphasizes communication but may not be aligning assessment methods with students' backgrounds.

John indicates that the absentee rate in his classes is very high—sometimes approaching 50 percent. How can he teach them if they don't come to class? He questions whether Katie is expecting too much when she wants her students to "explain" and "draw and label." After all, many students are not proficient with the basic skills. He claims that he keeps it simple by sticking strictly with the tests in the book, one test every Friday, with makeups on Monday. If students don't show up for the makeup, that can't be his fault, he argues.

The teacher tends to see the problem as the students' problem rather than as an instructional or curricular problem.

The teachers continue to discuss the problem. They recognize that although there are some things beyond their control, there are some things that they can control, for example, how they evaluate students. They also recognize that the cultural diversity of the students may require them to adjust how they have been testing and grading students. They decide on the following means of evaluating students:

The teachers decide to evaluate students on activities other than test and quiz scores.

1. *Journals.* Every student will keep a journal. The journal will count the same as a test grade. The daily entries will include examples worked out in class and various methods presented in class for solving problems. In addition, they will focus on the following items:

The teachers hope that the journal entries will increase students' disposition to do mathematics.

   a. What they learned that day.
   b. What they did in class that helped them learn.
   c. Why they think it is important to learn it.
   d. How they felt about the class that day.

The team felt that these questions would also keep them on their toes when preparing lessons. For example, question c will serve as a constant reminder to provide reasons why a topic is important to learn.

The teachers recognize that changing methods of evaluation has implications for their teaching as well.

2. *Class discussions.* Greater emphasis will be given to class discussions in evaluating students, thereby encouraging students to attend class. Students will be given more responsibility to present solutions and to explain procedures during class discussions.

The teachers will place a greater emphasis on students communicating mathematics.

The teachers decide to try an alternative method of giving quizzes.

3. *Quizzes.* All quizzes will be taken in pairs; students will be able to discuss their solutions with their partners. This will facilitate communication and help the students feel less tension when taking a quiz.

The teachers are making an effort to match assessment with students' backgrounds.

4. *Tests.* All tests, including makeup tests, will be shared among team members. They will maintain an emphasis on "explaining" and will try some open-ended items as well, but they will give the students greater latitude in responding. They will also make the tests shorter so that students will have more time to respond.

The evaluation of the program will be cyclical in nature.

The three teachers agree to reevaluate the effect of the program next December. Simone is impressed with what the teachers have done. She says that she will be interested in their assessment of the effectiveness of the program next December. She reminds them that the ultimate goal is to increase learning and to encourage more students to be successful in doing mathematics and to continue their study of mathematics. The high failure rate is counterproductive to achieving this goal. The teachers and the principal decide to meet informally once or twice a month to share ideas and make any modest revisions. They hope that by this time next year, they can point to the program as a model of success in increasing achievement and disposition to do mathematics. Simone is hopeful that the program will also help reduce the absentee rate among students.

The teachers and the principal will monitor the program's effectiveness.

STANDARD 8:
LEARNING ENVIRONMENTS

Assessment of the teacher's ability to create a learning environment that fosters the development of each students' mathematical power should provide evidence that the teacher—

♦ *conveys the notion that mathematics is a subject to be explored and created both individually and in collaboration with others;*

♦ *respects students and their ideas and encourages curiosity and spontaneity;*

♦ *encourages students to draw and validate their own conclusions;*

♦ *selects tasks that allow students to construct new meaning by building on and extending their prior knowledge;*

♦ *makes appropriate use of available resources;*

♦ *respects and responds to students' diverse interests and linguistic, cultural, and socioeconomic backgrounds in designing mathematical tasks;*

♦ *affirms and encourages full participation and continued study of mathematics by all students.*

Elaboration

Any assessment of the teaching of mathematics should consider whether the teacher has established and maintained an environment that promotes the development of students' mathematical power. A spirit of inquiry should pervade all mathematics teaching and learning. In establishing such an environment, the teacher must be sensitive to students' ideas and encourage mathematical communication among all students. Technology can be used effectively in creating such an environment, since it provides a tool for making mathematical explorations more efficient and accessible.

An inquiry-oriented classroom can be promoted by engaging students in extensive mathematical discussions and encouraging them to reason mathematically. The teacher must respect students' ideas by listening to them and incorporating their ideas into the class discussion. For their part, students should demonstrate a willingness to propose hypotheses, to support their own hypotheses, and support or challenge hypotheses set forth by the teacher or other students. As the teacher models encouragement and support for students and respects and accepts their ideas, so should students learn to support and respect each other and to work collaboratively and actively to solve problems and to validate proposed solutions.

Rather than rely solely on the logical structure of mathematics to determine and sequence classroom activities, the teacher should use his or her knowledge of students' understanding as a primary basis for selecting and sequencing mathematical tasks at an appropriate level of abstraction. Tasks that promote the active involvement of all students should be selected. It is imperative that the teacher set high expectations for every student and work vigorously to see that those expectations are met.

Ultimately, students must assume responsibility for their own learning. However, the teacher has the responsibility to create an environment in which students are encouraged to accept that responsibility. This standard identifies indicators for determining whether a teacher is creating such an environment.

Vignettes

8.1 A group of middle school teachers and Larry Parker, the district's mathematics coordinator, have been meeting regularly for the past semester. The group's goal is to improve the teaching and learning of various specific topics. They are now focusing on the concept of percent—a concept that is typically difficult for their students. Larry schedules his visits to individual classrooms so that he can observe lessons on percent. On Wednesday afternoon he observes Betty Mathison, an experienced sixth-grade teacher.

Betty has prepared transparencies of 100-square grids that are partially shaded. Each grid is shown briefly (a couple of seconds) on the overhead before she asks the question, "What percent of the grid has been shaded?"

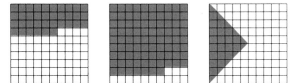

Examples of transparencies used.

Students write their guesses and their reasoning on a sheet of paper. Betty walks around the room and observes what they have written after showing each transparency. Betty then randomly selects students and asks them to indicate what their guesses are and to share their rationales for the guesses. The students begin to appreciate the strategies that their classmates are using.

Students give the following reasons for their estimates:

◆ The first one was 36% because I saw 3 rows of ten and 6 more.

◆ The first one was more than 30%. I saw three rows but didn't have time to count the other ones. It was probably about 35%.

◆ In the second one, I think there were 8 rows of ten and then 7 more. That makes 87%.

◆ I know there was one row that wasn't shaded so it had to be less than 90%. Maybe it was 86%.

◆ The last one was hard. I think it was about 20%. It seemed like it was one-fourth covered.

◆ The third one was less than 50% but I don't know exactly. Maybe it was 10%. I'm just guessing.

Betty is supportive of the students' responses; sometimes she asks them to elaborate on their reasons.

Following this whole-class activity, Betty has the students work in pairs to represent different percents using cardboard base-ten materials. In addition to asking students to represent whole number percents less than 100, Betty asks students how they might represent 150% or 200% or a percentage less than 1%, like 0.5%. Larry is impressed with the extension.

The supervisor and teachers work together to address problems related to the teaching and learning of percent.

The teacher selects a task that allows students to use what they have previously learned about fractions.

By having the students write their estimates, the teacher is involving all the students in the lesson.

The teacher is providing the students with an opportunity to share their mathematical reasoning.

The teacher encourages students to engage in mathematical reasoning.

The mathematics supervisor notes that the teacher provides students with an opportunity to represent different percents.

Betty is hoping the students will see the connection between representing percents like 36% and 87% to representing percents that are more than 100% or that are not whole number percents. She reminds the students of the earlier work they had done with decimal representations using the cardboard base-ten materials.

The teacher encourages students to make connections with previously learned material.

Both Larry and Betty are pleased with the lesson. Although the students had a great deal of difficulty representing percents greater than 100 and percents that are not whole numbers, they nevertheless felt that it was a good start in developing a concept of percent. Larry asks Betty to share the lesson with the other middle school teachers at next week's meeting. He wants the other teachers to see how Betty has made effective use of base-ten materials constructed from old cut-up cardboard boxes to teach percent. Larry indicates that he will try to secure funds so that the materials can be made from card stock and thus be more attractive for the students.

The supervisor is very supportive of the tasks the teacher designed.

The supervisor supports the teacher's creativity in using available resources.

8.2 Pierre Bordeaux, a master teacher, is on the telephone talking with his friend Sally Witt, a mathematics educator from Southwest State University. Sally would like her preservice teachers to observe a teacher who has a reputation for having students explore mathematics. She would also like her students to have the experience of evaluating a mathematics lesson. The methods students have developed criteria they think are important to the teaching of mathematics. In particular, they think the mathematical environment is important. Mr. Bordeaux indicates that he would welcome the university students and whatever comments they might have about his teaching. He suggests that the students come next Thursday because he is planning to do some explorations involving the Pythagorean theorem.

University methods students will confer with the teacher about their reactions to the lesson.

The mathematical environment and how it affects learning is recognized as being important.

The next Thursday Dr. Witt's methods class is sitting along the sides of Mr. Bordeaux's class ready to take notes. Mr. Bordeaux begins the class by asking his students if they remember anything about the Pythagorean theorem. The students offer that it has something to do with right triangles, the Greeks, and that it has a formula. Yvonne claims that it says $a^2 + b^2 = c^2$. Clare says that it works only for right triangles. Ben adds that the c side must be opposite the right angle. Mr. Bordeaux compliments them on remembering so much about the theorem and tells them that the Pythagorean theorem was also known to the ancient Egyptians and Chinese. He encourages them to check other books in the school library on the history of the Pythagorean theorem to see what else they can learn about it.

The teacher engages students in mathematical discourse.

The teacher demonstrates respect for the students and their cultural backgrounds. He also represents mathematics as a human endeavor.

Mr. Bordeaux asks the students to provide an interpretation or a drawing to represent what the theorem says. The students indicate that they don't know what he means. After some discussion, the students draw figures similar to the following and interpret the theorem in terms of the figures. They confess that they had thought of the theorem only in terms of squaring numbers.

The teacher asks the students to engage in mathematical communication.

The teacher helps the students extend their understanding of the theorem.

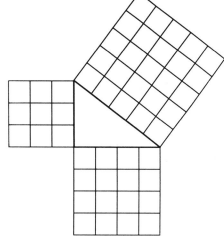

Mr. Bordeaux then asks them if other figures were drawn on the legs and the hypotenuse, would the

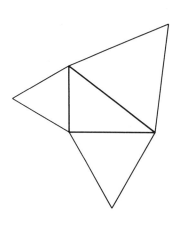

The teacher gives the students an opportunity to engage in mathematical exploration.

The university students take note of the teacher's questions.

The teacher is making appropriate use of resources.

The university students are impressed with how the teacher supports students in validating a conjecture.

The methods students focus on the exploring nature of the mathematical environment and on the emphasis given to mathematical communication.

The teacher poses a mathematical question.

The methods students observe that a different means of validating a conjecture is used.

The student forms a conjecture.

The university students observe that the teacher has asked a student to produce a counterexample.

same relationship among the areas hold true. Jeff asks, "Figures like what?" to which the teacher responds, "Like equilateral triangles." Both Mr. Bordeaux's students and Dr. Witt's students are struck by the question; they had never considered such an extension of the Pythagorean theorem. Most of the geometry students volunteer that it won't work; a few think it will. Having anticipated that there would be some disagreement, Mr. Bordeaux distributes a worksheet with three right triangles drawn, construction paper, and scissors. He instructs the students to work in pairs and construct and cut out equilateral triangles to fit on the sides of the right triangles.

Mr. Bordeaux encourages the students to cut the triangles into smaller pieces in order to determine if the sum of the areas of the two smaller equilateral triangles is equal to the area of the larger equilateral triangle. The university students observe the way the teacher engages the students.

The methods students are quite impressed with the mathematical activities. They observe that the students are busy exploring mathematics and that the teacher is supportive of their doing so. They are particularly impressed when Mr. Bordeaux has the students verbalize what they had found. This is consistent with the emphasis Dr. Witt had given in the methods class and with the criteria for effective teaching that they had established.

Mr. Bordeaux then asks the students to consider the case when semicircles are drawn on the sides of the triangles. The students draw the figure and determine that the sum of the areas of the semicircles on the two legs is equal to the area of the semicircle on the hypotenuse by using the formula for the area of the circle. Mr. Bordeaux points out that this is a different kind of justification from what they had previously used with the construction paper and the equilateral triangle. The methods students take note of this.

Nathaniel asks if the Pythagorean theorem works for any figures placed on the sides of the triangles. Manuel and David think not. They state that you can't just draw any rectangles on the sides—they would have to be related in some way. Mimi agrees. She says, "You couldn't expect one to be skinny and another one thick and expect it to work." The university students take note of Mr. Bordeaux's next question, "Can you draw an example to show one that doesn't work?" Mimi draws the following picture to illustrate her point:

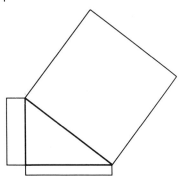

Mr. Bordeaux compliments Mimi for her counterexample. After vigorous discussion, the students decide to consider the case in which the lengths of the rectangles are twice as long as the widths. They will study this case as part of tomorrow's homework.

The methods students note that the teacher is supportive and observe that the geometry students take initiative in the discourse.

After the class, Mr. Bordeaux, Dr. Witt, and the university methods students meet to discuss the lesson. The methods students ask whether Mr. Bordeaux always uses materials like he did today to explore mathematics. He replies, "Not always," but indicates that he likes to use them as much as possible. He explains that it is very important for students to see mathematics as a subject to be explored and not just as statements in a textbook. He emphasizes to the university students that he likes to have the students work in pairs or in small groups to explore various mathematical topics. The methods students ask how he decided to pair the students. Mr. Bordeaux confesses that he hadn't thought much about it but that he would the next time. One of the university students recalled that one of the students had a broken hand and couldn't do much of the physical manipulation during the exploration. Mr. Bordeaux acknowledges this and indicates that he should have spent more time helping that student.

The university students inquire about the extent to which the teacher uses instructional aids in engaging students in exploration activities.

The university students ask how the teacher is meeting the needs of individual students by pairing them as he did.

The teacher acknowledges that he needs to consider more carefully how to organize the students.

Summary: Foci of Evaluation

The five standards in this section present foci for evaluating the teaching of mathematics. All these standards, and particularly Standard 6, emphasize the importance of significant mathematics when evaluating the teaching of mathematics. It is through encountering significant mathematics that students develop mathematical power. But attaining mathematical power requires more. It requires a disposition to do mathematics and an environment in which the processes of doing mathematics are continually emphasized.

All this cannot occur without teachers who present stimulating tasks and create an environment in which problem solving, reasoning, and communication are valued and promoted. Further, the message teachers send students should not be limited to instruction alone; it must also include what and how mathematical learning is assessed. It is through assessment that we communicate to our students what mathematical outcomes are valued.

A consistent message throughout the standards for the evaluation of teaching is the importance of teachers' being reflective about their teaching and working with colleagues and supervisors to improve their teaching. Although the standards in this section can provide a focus for improvement, such improvement is more likely to occur when teachers have the support to engage in professional development. As suggested in vignette 8.2, professional development spans a teacher's professional life. In the next section standards for the professional development of teachers are presented and discussed.

STANDARDS FOR THE PROFESSIONAL DEVELOPMENT OF TEACHERS OF MATHEMATICS

OVERVIEW

This section presents six standards for the professional development of teachers of mathematics:

1. Experiencing Good Mathematics Teaching

2. Knowing Mathematics and School Mathematics

3. Knowing Students as Learners of Mathematics

4. Knowing Mathematical Pedagogy

5. Developing as a Teacher of Mathematics

6. The Teacher's Role in Professional Development

INTRODUCTION

Teaching mathematics is a complex endeavor. It demands knowledge of mathematics, students, and teaching as well as opportunities to apply this knowledge in a variety of field-based settings. It requires an understanding of the impact that socioeconomic background, cultural heritage, attitudes and beliefs, and political climate have on the learning environment. Above all, it entails developing a personal knowledge of oneself that combines sensitivity and responsiveness to learners with the knowledge, skills, understandings, and dispositions to teach mathematics.

The standards for the professional development of teachers of mathematics address the needs of preservice and in-service teachers of mathematics at the K–12 levels. These standards apply to introductory programs that prepare teachers of mathematics; programs that provide advanced study for teachers of mathematics; and various continuing education seminars, workshops, and other learning experiences in which teachers of mathematics participate throughout their careers.

Those who teach mathematics bring with them experiences as learners in mathematics classes from elementary school through their college and university careers. These experiences influence the ways they think about the teaching process, their choice of teaching as a career, and the subsequent ways in which they are involved with professional development programs (Ball 1988).

In the early stages of their careers, preservice teachers of mathematics are often involved in developing their knowledge, skills, understandings, and dispositions to teach mathematics. This development includes knowledge of mathematics, students, and teaching and involves having opportunities to integrate and apply this learning as practitioners. During this time, they are involved in a number of clinical experiences in which they interact with supervising or experienced teachers who function as mentors. This period is one in which new ideas can be tried, analyzed, and questioned with support and encouragement.

The first few years of teaching present a very different period in the professional development process. Initial teaching assignments and support structures play a significant role in shaping beginning teachers' views of the profession and their commitments to it. The focus is on being a teacher—planning for instruction, managing students' learning, responding to the changing needs of the learning environment—and involves a comprehensive application of what teachers have learned and experienced as part of their preservice programs. New issues are confronted, and knowledge and skills are built daily, more often within the context of the teaching environment than through formal continuing education.

As teachers of mathematics become more experienced, collegial interactions increase and teachers assume a different role. Experienced teachers of mathematics become more involved in decisions about curriculum and staff development programs. Indeed, the identification of staff development needs and the development of programs to meet these needs can become part of the professional responsibilities of an experienced teacher. In addition, experienced teachers may become mentors to beginning or developing teachers at this time in their careers. There is an ebb and flow to their need for more formal continuing education. Their ability to engage in ongoing analysis of their own teaching and learning is often central to their seeking experiences that address knowing mathematics, knowing students, and knowing teaching.

ASSUMPTIONS

Several basic assumptions provide the foundation for the standards detailed in this section:

1. *The* Curriculum and Evaluation Standards for School Mathematics *provides the vision of mathematics education at the K–12 level that is the basis for the professional development standards.* The Curriculum and Evaluation Standards, designed to create a coherent vision of what it means to be mathematically powerful in today's world, provides an in depth focus on what is valued in the learning of mathematics K–12 and in the evaluation of that learning. Teachers must have a thorough understanding of the message in this document. Their education should include the development of the knowledge, skills, understandings, and dispositions needed to implement the recommended standards.

2. *Teachers are influenced by the teaching they see and experience.* Teachers' own experiences have a profound impact on their knowledge of, beliefs about, and attitudes toward mathematics, students, and teaching. Teachers' thirteen years as learners of K–12 mathematics provide them with images and models—conscious or unconscious—of what it means to teach and learn mathematics. They add to this many other learning experiences such as formal college preparation, clinical and field-based observation and practice teaching, the influence of school culture and colleagues within their immediate teaching environments, and in-service and advanced educational experiences. All these experiences convey messages about what constitutes appropriate teaching and learning. Such powerful influences need to be addressed when helping teachers learn to teach in new ways.

3. *Learning to teach is a process of integration.* Although the standards for the professional development of teachers of mathematics address various components of teacher knowledge and practice separately, the final success for any teacher is the integration of theory and practice. Ideally, teachers should engage simultaneously in studies of mathematics

content and mathematics pedagogy. Teachers should be able to comment and reflect on their own learning environments at the same time they are involved in clinical and field-based teaching experiences. As different instructional strategies are implemented, teachers should be discussing the research that supports their choices of these strategies. Such integration is not easily achieved. However, it is a goal to strive for as reform of teacher education is pursued.

4. *The education of teachers of mathematics is an ongoing process.* Teachers are in a constant state of "becoming." Being a teacher implies a dynamic and continuous process of growth that spans a career. Teachers' growth requires commitment to professional development aimed at improving their teaching on the basis of increased experience, new knowledge, and awareness of educational reforms. Their growth is deeply embedded in their philosophies of learning, their attitudes and beliefs about learners and mathematics, and their willingness to make changes in how and what they teach. Their growth is also affected by numerous external agents including school administrators, educational policymakers, college and university faculty, parents, and the students themselves.

5. *There are level-specific needs for the education of teachers of mathematics.* Although certain knowledge, skills, and abilities are common to the preservice and continuing education of all teachers of mathematics, level-specific needs—such as the learning needs of elementary, middle, and secondary school students—must be addressed. For example, in considering the development of teachers' knowledge of mathematics, different expectations about the breadth, depth, and scope of the content knowledge are needed for teaching at each grade level. These different needs must be addressed when considering teachers' knowledge of students and teaching.

ALL STUDENTS

Throughout these standards the phrase *all students* is used often. By this phrase we mean to set the mathematical education of every child as the goal for mathematics teaching at all levels, K–12. In April 1990, the NCTM Board of Directors endorsed the following statement:

> As a professional organization and as individuals within that organization, the Board of Directors sees the comprehensive mathematics education of every child as its most compelling goal.
>
> By "every child" we mean specifically—
>
> - students who have been denied access in any way to educational opportunities as well as those who have not;
> - students who are African American, Hispanic, American Indian, and other minorities as well as those who are considered to be a part of the majority;
> - students who are female as well as those who are male; and
> - students who have not been successful in school and in mathematics as well as those who have been successful.

It is essential that schools and communities accept the goal of mathematical education for every child. However, this does not mean that every child will have the same interests or capabilities in mathematics. It does mean that we will have to examine our fundamental expectations about what children can learn and do and that we will have to strive to create learning environments in which raised expectations for children can be met.

ORGANIZATION

In deciding how to present and elaborate the ideas underlying each of the six standards in this section, we faced two dilemmas. First, these standards seek to answer the question: What is the nature of the professional development experiences and opportunities needed in order for teachers to teach mathematics as described in the first section, "Standards for Teaching Mathematics"? Depending on one's view, this is a deceptively simple or an obviously complex question. In our efforts to be succinct yet comprehensive, we struggled with both the breadth and depth needed to provide a map for those interested in possible paths to answer this question. There are a variety of perspectives that must be considered in identifying what constitutes standards for the professional development of teachers of mathematics. We believe these standards help the professional community address many of these perspectives.

Our second dilemma evolved as we struggled to balance our message with that of the first section. The distinctions between the two sections are not always clear; they are intertwined in mutually supportive ways. Consequently, we chose not to address directly what is meant by tasks, discourse, environment, and analysis, although we make frequent reference to these ideas and integrate them throughout these standards. We urge readers of this section to carefully review the first section as part of their efforts to consider what constitutes appropriate opportunities for professional development.

The statement of each of the six standards is first elaborated with an explanation of its main ideas and occasionally highlighted by quotations from mathematicians, mathematics educators, teachers, and students. Then, for each standard, we follow with annotated vignettes that show and extend these ideas through a variety of contexts related to the preservice and continuing education of teachers of mathematics. Drawn from transcripts, observations, and experiences, the vignettes are selected to illustrate a range of professional development opportunities and issues. The commentaries in the outside column focus on issues pertinent to that standard and, in some instances, include additional detail that elaborates as well as annotates.

SUMMARY

The standards in this section focus on what we believe are essential components for the professional development of teachers of mathematics. They comprise the threads that are woven as the fabric of successful mathematics teaching: personal experiences in contexts that model and value good mathematics teaching; ongoing development of knowledge about mathematics, students, and teaching; numerous and diverse opportunities to apply knowledge and experience through practice; and the gradual assumption of responsibilities for professional growth and change. Ideally, the weave of the fabric will evolve and change, reflecting the numerous stages to be explored in the career-long development of mathematics teachers.

STANDARD 1:
EXPERIENCING GOOD MATHEMATICS TEACHING

Mathematics and mathematics education instructors in preservice and continuing education programs should model good mathematics teaching by—

♦ *posing worthwhile mathematical tasks;*

♦ *engaging teachers in mathematical discourse;*

♦ *enhancing mathematical discourse through the use of a variety of tools, including calculators, computers, and physical and pictorial models;*

♦ *creating learning environments that support and encourage mathematical reasoning and teachers' dispositions and abilities to do mathematics;*

♦ *expecting and encouraging teachers to take intellectual risks in doing mathematics and to work independently and collaboratively;*

♦ *representing mathematics as an ongoing human activity;*

♦ *affirming and supporting full participation and continued study of mathematics by all students.*

Elaboration

I think that an investigative nature is essential. We can't make students into seekers if we aren't seekers ourselves. (A university mathematician)

The experiences that mathematics teachers have while learning mathematics have a powerful impact on the education they provide their students. Prospective and practicing teachers spend many hours in mathematics and mathematics education classes, workshops, seminars, and other structured learning environments. Through these experiences, they develop ideas about what it means to teach mathematics, beliefs about successful and unsuccessful classroom practices, and strategies and techniques for teaching particular topics. Those from whom they are learning are role models who contribute to an evolving vision of what mathematics is and how mathematics is learned.

My students were passive and docile. I felt I was perpetuating those attitudes in them. I wanted to devise a non-lecture teaching format, but I wasn't trained for such a search. They don't teach pedagogical techniques in mathematics graduate school. (Macrorie 1984, p. 66)

Instructors of mathematics and mathematics education in any and all learning situations need to address the major components of teaching— tasks, discourse, environment, and analysis of teaching—as detailed in section 1 of this document, "Standards for Teaching Mathematics." This vision of teaching redirects mathematics instruction from a focus on presenting content through lecture and demonstration to a focus on active participation and involvement. Mathematics instructors do not simply "deliver" content; rather, they facilitate learners' construction of their own knowledge of mathematics. Sometimes they stand back, letting students puzzle and come up with their own solutions. Sometimes they push and lead, helping students to reach particular sensible conclusions. And sometimes they help students by modeling or telling. Mathematics instructors do this through their choice of tasks and tools for instruction and their attention to the nature of the mathematical discourse that occurs in the learning environments.

> I was once questioned by a colleague, " But you seem to imply that you are not doing much lecturing. Isn't lecturing necessary to cover the content?" You know, it really isn't anymore, and it doesn't seem to be really important to my students. (A university mathematician)

Mathematics and mathematics education instruction should enable all learners to experience mathematics as a dynamic engagement in solving problems. These experiences should be designed deliberately to help teachers rethink their conceptions of what mathematics is, what a mathematics class is like, and how mathematics is learned. Instruction should be organized around searching for solutions to problems and should include continuing opportunities to talk about mathematics. Working in groups is an excellent way for learners to explore, develop mathematical arguments, conjecture, validate possible solutions, and identify connections among mathematical ideas. In such experiences, teachers should be encouraged to generalize solutions and communicate results from their explorations of mathematical ideas visually, in writing, or through dialogue and discussion.

Representations are crucial to the development of mathematical thinking, and through their use, mathematical ideas can be modeled, important relationships identified and clarified, and understandings fostered. Physical models, materials, calculators, and computers help provide the array of rich and substantive experiences needed to build a deep and comprehensive knowledge of mathematical concepts and procedures. The experiences teachers have in these learning environments form expectations—implicitly or explicitly—of what constitutes good mathematics instruction. Such experiences provide the core from which teachers will eventually build learning environments for their own students.

> For a long time my undergraduate courses in differential geometry have profited from slides and films and videotapes, but I wasn't prepared for the tremendous advance that came when the students were able to work interactively with computers. (Banchoff 1986, pp. 8–9)

Such instruction requires substantial changes in the philosophy and strategies of mathematics and mathematics education instructors at the college level and beyond who are involved in the preservice and continuing education of teachers of mathematics. Instructors need to experiment with new tasks, tools, and modes of classroom interaction and share and model new instructional strategies. This necessitates collegial interaction and support, as well as participation in professional development opportunities. Similarly, such changes necessitate changing the recognition and reward systems in colleges and universities. Also, school districts need to revise their perspectives on the kind of in-service support needed to effect substantive change. Finally, such changes place new expectations on teachers as students in their participation and engagement in learning. This challenges mathematics and mathematics education instructors to foster changes in their students' preconceived and generally traditional views about the way learning occurs.

Vignettes

1.1 Prospective middle school teachers are coming to the end of a yearlong mathematics course. By now they take it for granted that they are expected to make sense of mathematics, develop their own problems, make connections, and come up with further questions to extend their thinking. This hasn't happened automatically; the mathematics professor has worked hard to achieve her goal of encouraging students

The instructor's goals for her students require changes in her own teaching—she has worked to help them become more active participants in, and creators of, their own learning.

to develop greater reliance on their abilities to make sense of mathematics.

"It hasn't been easy to shift these students' expectations from wanting answers from the instructor to a point where they accept, and in fact demand, that they have a chance to make sense of a situation themselves."

Early in the year, a student who was interviewed about the class noted, "Up until the university placement exam I just plugged numbers in and always got good grades. It had been a long time since I had math. I couldn't remember the way to do lots of the problems or appropriate formulas. I had no ability to tackle problems if I didn't know the formula. I like to plug numbers into formulas. This math class is very upsetting. This is the first time I ever thought about why. In high school algebra we just plugged in the numbers...just waited for the formulas. I realize I am going to have to learn to think about it [why] if I expect to teach math."

Now, later in the year, students are working on a research project that allows them to apply and extend the ideas they been have studying in probability and statistics. They have been challenged by their instructor to design and carry out an investigation to better understand some situation that they find interesting. The class decided that they wanted to know who Mr. and Ms. Typical Student were on their campus. They have spent several days designing questionnaires and gathering information.

"Oh, no! Now we don't know which data we have already entered! We will have to start all over again. We'd better get ourselves organized this time." From another group, "Would you like to know how we are keeping track of our data?"

"This is looking good! Would you look at that! Students who are working seem to have higher grade-point averages. Let's do a graph of grades versus hours of work each week and see if there is a relationship here. Let's go to the computer to do this."

Students find this kind of involvement in the class different from other experiences they have had in mathematics. One of the students pointed out a contrast between problems in this class and assignments in a typical math class:

"I still get frustrated a lot, but I am more satisfied when I can figure things out rather than just doing busy work. I used to think that math was just busy work. Do thirty problems on the distributive property when they're just different numbers...all the same thing, mindless by the thirtieth problem. Here's thirty problems, do them. Turn them in tomorrow. Here's thirty more problems. This class doesn't have any busy work."

1.2 Ron Adams, professor of mathematics at a large university, uses computer numerical, graphic, and symbol manipulation programs to support a new calculus course that emphasizes concepts, principles, and applications of the subject. He engages students in active exploration and discussion of these ideas while they learn. He plans his initial class sessions to establish these directions.

In his overview comments, Ron suggests to his students that calculus provides an array of methods for studying relations among numerical variables. As an example, he identifies three pairs of related variables:

The instructor acknowledges that change is difficult and requires deliberate efforts to encourage students to think differently about the ways they approach mathematics.

Interviews with students provide a way for the instructor to hear what students are thinking about mathematics and their own learning.

Students confront their own perceptions about what it means to learn mathematics. Their level of discomfort is an indication that change is happening.

The instructor has identified long-term projects as worthwhile tasks. In this instance, students have been able to define their own investigation, making it relevant to their own lives and interests.

Students must define the task, work independently and collaboratively to accomplish their work, and use a variety of tools.

The students have calculators and computers to help with graphing and statistical analyses. These tools encourage conjecturing and "what if-ing" as they investigate the data.

A different environment has been created that supports and encourages mathematical reasoning. Such an environment influences students' dispositions to do mathematics. For these prospective teachers, it can have an impact on how they think about their own teaching.

The instructor explores ways to enhance mathematical discourse through the use of technology.

The instructor helps students realize that mathematics is useful in understanding a variety of real-world relationships.

Creating a learning environment that encourages students to think for themselves, not simply copy and then memorize the words of a lecture, is critical in helping them learn calculus and also in providing a model of how they, in turn, can teach others.

SAT scores and freshman GPAs
Oil supplies and gasoline prices
Rainfall and mosquito populations

He encourages students, working with partners, to discuss each of these situations and then share their ideas about how the variables in each pair may relate to each other.

The instructor uses multiple representations of mathematical ideas to help students with different learning styles understand key ideas, recognizing that students' ability to connect different representations of the same information is a critical test of deep understanding.

Following this, Ron introduces methods for representing quantitative relations that can help in identifying the critical properties of those relations. He selects two other pairs of related variables:

A table of student absences from school during a flu epidemic—data are reported by days of the week.

A graph of the growth of the fish population in a lake—data are shown in units of time.

The instructor encourages discourse, providing questions that will help students think about the mathematics of the situations.

Working in small groups of three to four students, they analyze each representation, noting interesting individual facts and global patterns. Ron distributes a set of questions to help students in their obervations, guiding them to notice key issues like rates of change, average values, maximum or minimum points, and representations of those concepts in tables of values and in shapes of graphs.

The instructor focuses on a developmental sequence that involves interpreting concrete tasks within familiar settings and moves to tasks that provide less relevant contexts and include constructing representations of relations through tables and graphs.

Following a discussion of observations, Ron extends this awareness activity, asking students to construct tables of values and graphs that they believe represent the shapes of likely relations between the depth of tidal water and the time of day or between the profit for a resort hotel and the average price charged for rooms.

Ron encourages the students, working again in small groups, to talk about such relations and their representations, avoiding an emphasis on exact "right" answers.

The instructor does show his students how to perform some useful procedures. However, the purpose is to help students become quickly involved in using the computer tools they need to explore mathematics.

Ron moves to a third representation, symbolic rules, that can be used to study significant properties of functions. He demonstrates how computer programs use algebraic rules for functions to produce tables of values and graphs and then to identify key points and overall properties of those functions. Using examples of familiar situations, Ron shows students how to produce computer tables and graphs for the following:

A linear demand function and the related quadratic revenue function.

The periodic functions describing voltage in an alternating current circuit.

An exponential population growth function.

The instructor participates with the students in the investigations.

In each case, he and the students work together to describe the important properties and explore how the function is represented both numerically and graphically.

With early access to technology, mathematics can truly become a laboratory subject.

As part of a computer lab project, student teams work together to develop skill using numerical and graphic tools and understanding the meaning of the results.

Students are engaged in developing the mathematical theory that emerges from their concrete experiences.

Throughout the course, the use of computer tools allows students to examine an impressive array of function types and to tackle problems of convincingly realistic complexity without stopping to develop proficiency in symbolic calculations. After exploring how ideas arise in realistic contexts, Ron and his students work together to construct the theoretical framework of those ideas out of their shared experiences.

1.3 Rita Jamison, an associate professor in the mathematics department, has been participating in a project that teams a mathematician with a mathematics educator to help introduce both of them to the use of technology and to an interactive approach to teaching undergraduate mathematics classes taken by preservice secondary mathematics teachers. Talking with the project director, Tom Bethel, she reflects on her experiences teaming with Bill Moorehead, a member of the mathematics education faculty.

Rita: I don't believe that I'll ever go back to my old ways of teaching undergraduates. I used to spend most of my time presenting formulas and going over homework. My students used to repeat back what I taught, always seeking the quick rule. Now, after a quick introduction to a new topic, Bill and I focus on problem solving. Students are really thinking; they create their own problems, explore extensions and elaborations, and are willing to do a lot of "what if" thinking. What helps is both the way I have changed my teaching in general and the use of the graphing calculator as an instructional tool.

Tom: Bill stopped me in the hall the other day. He indicated that you two had worked together quite well throughout the semester.

Rita: We really did. Bill and I decided to use graphing calculators throughout the course in order to reduce the emphasis on manipulation and computation. The department bought calculators so that we could loan one to each student in the class. We helped the students learn how to use the calculators during class and for all assignments and tests. It was much easier to do than I had imagined.

Bill: Hello, Rita. Hey, Tom. Are you talking about our work this semester?

Rita: Yes. I was just going to say that what really helped was spending a great deal of time rethinking how to present material so that students were more involved. Sometimes you sat in the back of the room and, after class, shared your observations about student participation.

Bill: And at other times we team taught a lesson. And sometimes I would teach, and Rita observed.

Rita: Yes, this all helped me become much more reflective about my teaching. I began to focus on ways to ask more engaging questions.

Bill: We also used cooperative learning strategies and relied on a great deal of class discussion to promote talking and thinking about mathematics. Rita's students became quite involved, bringing in problems from their own experience. For example, one student said her parents were willing to loan her $8000 at 9% interest for her to purchase a car. She explained that her parents only wanted her to pay them $50 each month. After some work with their calculators, other students told her that she would never repay the loan with that size payment!"

Rita: I was worried that I might not be able to cover the syllabus because of the extra time required to teach students how to use the calculator and the time needed for students to figure things out and discuss and work. Instead, we actually got more done. Taking the time to help students make sense of the mathematics enabled them to make connections between topics. New topics seemed more logical.

Tom: The changes you and Bill have made in your teaching are affecting the preservice teachers in the mathematics courses we are teaching.

Rita: Even other faculty! After looking at the midterm exam for this pilot course, Dave Smith commented to me, "Your students can do this?" He and I actually are planning to work together next year.

The structure and support system at the university level has been changed to accommodate interdepartmental collaboration.

Students are involved in worthwhile mathematical tasks that move them beyond skills. Technology is used to enhance mathematical discourse. Preservice teachers are able to independently explore their own conjectures and to extend their mathematical ideas.

University faculty support each other in changing mathematics curriculum and instructional strategies, including the use of calculators to enhance the classroom discourse.

Faculty explore ways to encourage mathematical discourse in the classroom. Together they try various options.

Through posing worthwhile tasks and carefully selecting questions to facilitate discourse, the instructor encourages students to reason mathematically and take intellectual risks. In this environment, they begin to make connections between the mathematics they are learning and life applications. They are disposed to reason about real-ife problems and investigate their solutions.

The instructor has support from her institution for trying changes in the way her students do mathematics.

As some faculty change and experience success, others become interested and consider similar changes themselves.

STANDARD 2:
KNOWING MATHEMATICS AND
SCHOOL MATHEMATICS

The education of teachers of mathematics should develop their knowledge of the content and discourse of mathematics, including—

♦ *mathematical concepts and procedures and the connections among them;*

♦ *multiple representations of mathematical concepts and procedures;*

♦ *ways to reason mathematically, solve problems, and communicate mathematics effectively at different levels of formality;*

and, in addition, develop their perspectives on—

♦ *the nature of mathematics, the contributions of different cultures toward the development of mathematics, and the role of mathematics in culture and society;*

♦ *the changes in the nature of mathematics and the way we teach, learn, and do mathematics resulting from the availability of technology;*

♦ *school mathematics within the discipline of mathematics;*

♦ *the changing nature of school mathematics, its relationships to other school subjects, and its applications in society.*

Elaboration

Knowledge of both the content and discourse of mathematics is an essential component of teachers' preparation for the profession. Teachers' comfort with, and confidence in, their own knowledge of mathematics affects both what they teach and how they teach it. Their conceptions of mathematics shape their choice of worthwhile mathematical tasks, the kinds of learning environments they create, and the discourse in their classrooms.

Knowing mathematics includes understanding specific concepts and procedures as well as the process of doing mathematics. Mathematics involves the study of concepts and properties of numbers, geometric objects, functions, and their uses—identifying, counting, measuring, comparing, locating, describing, constructing, transforming, and modeling. The relationships and recurring patterns among these objects and the operations on these objects lead to the building of such mathematical structures as number systems, groups, or vector spaces and the study of the similarities and differences among these structures. Mathematical concepts and structural properties are used to create powerful algorithms or procedures for solving whole classes of problems. At any level of mathematical study, there are important and appropriate concepts and procedures to be studied.

Such knowledge ought not to be developed in isolation. The ability to identify, define, and discuss concepts and procedures, to develop an understanding of the connections among them, and to appreciate the relationship of mathematics to other fields is critically important. Mathematics both arises out of, and influences continued development of, other fields. Advances in mathematical thought spur advances in physics. Advances in computer science raise new mathematical problems to be solved.

> Somebody once quoted a student saying, "The reason I didn't go into mathematics is because my instructors were never interested in hearing what I thought. It was always what they thought." I think that interactions with students are absolutely critical. Interaction, engagement, listening to students, being co-learners with students...these are important aspects. (A university mathematician)

Knowing mathematics also involves the larger context of mathematical discourse in which specific concepts and procedures are embedded. Discourse in mathematics centers on examining patterns, abstracting, generalizing, and making convincing mathematical arguments. It involves the role of definitions, examples, and counterexamples and the use of assumptions, evidence, and proof. Framing mathematical questions and conjectures, constructing and evaluating arguments, making connections, and communicating mathematical ideas all are important aspects of mathematical discourse. Engaging in mathematical discourse is central to how teachers come to know mathematics; to develop confidence in their own abilities to do mathematics; and to become aware of, and have an appreciation for, the place of discourse in the discipline of mathematics.

> I do think a rigorous proof can be worked out by a group of students reasoning together. One student may pick out a nuance of a problem that triggers the key to the next step for another. Students can also learn there are different methods of approaching the same problem by working together. (Gilligan, Lyons, and Hanmer 1990, p. 295)

As part of the environment of discourse, the development of abilities in mathematical reasoning and problem solving are essential. Mathematical reasoning involves an interplay between intuitive, informal exploration and formal, systematic proof. All too often, the formal written record of mathematics is what teachers study. The struggles, the false starts, and the informal investigations that lead to an elegant proof frequently are missing. Teachers need opportunities to construct mathematics for themselves and not just experience the record of others' constructions. In addition, teachers need to interact with others to pose and solve problems in order to develop a repertoire of problem-solving strategies.

> All my life I have known that I learned math by doing homework with my friends, comparing answers on the telephone, in the dorms, or on the way to school. This included male and female friends in high school and classmates in an all-female college. It was never something we would admit. If anyone ever found out we hadn't "done our own work," we felt wrong and accused of having cheated. Yet all real intellectual pursuits and learning take place with exchanges of information and ideas. We do not learn in a vacuum. There is as much learning that takes place in the small groups of two or three as there is that takes place at the individual desk. (Gilligan, Lyons, and Hanmer 1990, p. 294)

As an ongoing product of human activity, mathematics is a dynamic and expanding system of connected principles and ideas constructed through exploration and investigation. Developing such a perspective includes an appreciation for the historical and cultural contributions made to the development of mathematics. It provides a provocative backdrop that may be useful in motivating students as they approach new subject matter and in encouraging the full participation and continued study of mathematics by all students.

When I taught Functional Analysis I tried to put it into a historical context. Why did people look at this question or that question? Well, historically it seems that very abstract questions tend to come out of questions that are somewhat less abstract, and those come from questions even less abstract, until they finally get back to something that leads the students to say, "Well, that's kind of a natural question, isn't it?" (Macrorie 1984, pp. 65–66)

Mathematics is a dynamic discipline that continues to grow and expand in its uses in our culture. Teachers will be called on to adapt to curriculum changes that this growth will entail. The study of some of the contributions made to the development of mathematics by different cultures should provide teachers with resources to use in motivating students as they approach new subject matter.

More and more mathematicians of all sorts find themselves inspired by phenomena that arise from computer graphics. They visit laboratories, collaborating with computer scientists and with students who take for granted subject matter which simply did not exist a generation ago. (Banchoff 1986, p. 10)

Technology is a vital force in learning, teaching, and doing mathematics, providing new approaches for solving problems and influencing the kinds of questions that are investigated. It should play a significant role in the teaching and learning of mathematics. There are a variety of ways technology may be used to enhance and extend mathematics learning and teaching. By far the most promising are in the areas of problem posing and problem solving in activities that permit students to design their own explorations and create their own mathematics.

Technology changes the nature and emphasis of the content of mathematics as well as the pedagogical strategies used to teach mathematics. Indeed, one central issue revolves around the fact that some of the computational procedures that have formed the basis for mathematics courses at all levels are no longer essential. Performing computational and representational procedures by hand is time-consuming, and students often lose sight of mathematical insights or discoveries as they become mired in the mechanics of producing the results. With the introduction of technology, it is possible to de-emphasize algorithmic skills; the resulting void may be filled by an increased emphasis on the development of mathematical concepts. Technology—computers and calculators—saves time and, more important, gives students access to powerful new ways to explore concepts at a depth that has not been possible in the past.

Central to the preparation for teaching mathematics is the development of a deep understanding of the mathematics of the school curriculum and how it fits within the discipline of mathematics. Too often, it is taken for granted that teachers' knowledge of the content of school mathematics is in place by the time they complete their own K–12 learning experiences. Teachers need opportunities to revisit school mathematics topics in ways that will allow them to develop deeper understandings of the subtle ideas and relationships that are involved between and among concepts.

Such opportunities should include developing broad understandings of significant mathematics concepts and how they are related to other parts of the curriculum. This includes opportunities to develop a substantial overview of the mathematics curriculum. At all levels, teachers need to see the "big" picture of mathematics across the elementary, middle, and high school years. To use a geographic analogy, teachers need to have a mental roadmap that shows the major cities (curriculum topics) and the roads (mathematical connections) among them. Such a mathematical map should also highlight the importance of connections between mathematics and other school subjects and between mathematics and situations in nonschool settings out of which mathematics arises or in which it is applied.

Common Experiences in the Mathematical Education of Teachers

There are common experiences that should be ingredients in the ways teachers of mathematics build and extend their knowledge of mathematics. Regardless of the context, the following themes, as suggested in the *Curriculum and Evaluation Standards for School Mathematics,* should be prominent in these experiences:

- ◆ Problem solving in mathematics
- ◆ Communication in mathematics
- ◆ Reasoning in mathematics
- ◆ Mathematical connections (both within the discipline and to its uses in the world around us)

In addition, mathematical experiences for all teachers should foster—

- ◆ the disposition to do mathematics;
- ◆ the confidence to learn mathematics independently;
- ◆ the development and application of mathematical language and symbolism;
- ◆ a view of mathematics as a study of patterns and relationships;
- ◆ perspectives on the nature of mathematics through a historical and cultural approach.

These experiences may occur in mathematics courses, workshops, conferences, or other professional development activities. In the process of constructing and developing these experiences, appropriate attention to, and use of, mathematical modeling and technology should be included to enhance the teaching and learning of the mathematical ideas. To this end, teachers should become familiar with instructional technologies that provide powerful numerical, symbolic, and graphical tools for the exploration, investigation, and application of mathematics. These technologies should be incorporated in instruction and used for assignments whenever such inclusion is feasible and can add to student insight and understanding.

The discussion that follows identifies the mathematics content needed by all teachers in grades K–12, the additional mathematics needed by teachers in grades 5–8 and 9–12, and finally the additional mathematics needed by those who plan to teach mathematics in grades 9–12. This ensures that teachers at all grade levels have not only a thorough understanding of the mathematics they are teaching but also a vision of where that mathematics is leading.

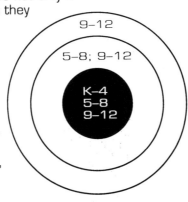

Mathematics for All Teachers in Grades K–4, 5–8, and 9–12

Foundational knowledge in mathematics is essential both for those teaching mathematics at grades K–4 and for those teaching mathematics at grades 5–8 and 9–12. With regard to specific content preparation, the mathematical education of all K–12 teachers should include the following:

Number systems and number sense. Teachers of mathematics should have a well-developed number sense (including mental mathematics, estimation, and reasonableness of results) and an understanding of the use of number concepts, operations, and properties (including basic number theory), of the role of algorithms, and of place value. In setting the view of these ideas in the curriculum, teachers should be able to extend the number systems from the whole numbers to fractions and integers, then rationals and real numbers, including a discussion of the extension of the operations, properties, and ordering. Notions of fractions, decimals, percents, ratio, and proportion should be developed through problems with an applied flavor.

Geometry. Young students have an informal and intuitive idea of size and shape. Teachers need to build on this informal background in the area of geometry. Teachers of mathematics should understand how geometry is used to describe the world in which we live and how geometry can be used to solve real-world problems. Analysis of two- and three-dimensional figures should include the study of tessellations, symmetry, polygons, polyhedra, and curved shapes. Synthetic, coordinate, and transformational geometry should be used to provide opportunities for teachers to solve problems and to hone their skills in building justifications and coherent arguments for the plausibility of conjectures. Throughout the experience, spatial visualization should be emphasized.

Measurement. The concept of measurement needs to be understood from the perspective of its historical development. The attributes of what we measure include length, area, volume, capacity, time, temperature, angles, weight, and mass. Teachers should understand that the units to record measure are different from the process of measurement itself. These ideas should be reinforced through varied experiences, using both standard and nonstandard units where students learn to estimate lengths, areas, and so on. Of particular importance should be an understanding of the Système International d'Unités (the metric system). Derivations of the formulas for the perimeter, area, and volume of common figures should be approached through meaningful explorations. Indirect measurement and its many applications should be studied.

Statistics and probability. Teachers should have a variety of experiences in the collection, organization, representation, analysis, and interpretation of data. Key statistical concepts for all teachers include measures of central tendency, measures of variation (range, standard deviation, interquartile range, and outliers), and general distributions. Representations of data should include various types of graphs, including bar, line, circle, and pictographs as well as line plots, stem-and-leaf plots, box plots, histograms, and scatter plots. Probability of simple and compound events and its use in quantifying uncertainty should be built into these experiences. Students should have opportunities to explore empirical probability from simulations and from data they have collected and to analyze theoretical probability on the basis of a description of the underlying sample space. Probability trees and simulations using objects such as spinners, dice, slips of paper, and so on should be used to solve problems.

Functions and use of variables. Teachers need to experience the development of mathematical language and symbolism and how these have influenced the way we communicate mathematical ideas. Also, experience in representing and solving problems requiring the use of variables is important. To build bridges for their students to the mathematics that comes later in the school curriculum, teachers must have a basic understanding of the concepts of functions and their use in the growth of mathematical ideas. Understanding different representations of functions (tabular, graphical, symbolic, verbal), how to move among these representations, and the strengths and limitations of each is fundamental. The distinction between continuous and discrete approaches in the solution of mathematical problems should also be a part of the experiences provided for these teachers and should be introduced initially at an intuitive and informal level.

Additional Mathematics for Teachers in Grades 5–8 and 9–12

Teachers of mathematics at grades 5–8 and 9–12 must present mathematics that builds on the students' background established in the elementary grades. New mathematical knowledge should deepen the understandings of the topics already noted and introduce new and worthwhile mathematical ideas. With regard to specific content preparation, the mathematical education of teachers at the 5–12 level should include and extend the material described earlier by including the following:

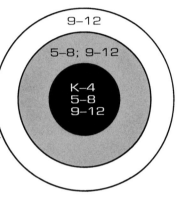

Number systems and algebraic structures. The system of real numbers should be extended to the complex numbers. Investigations of selected algebraic structures should include concrete examples such as clock arithmetic, modular systems, and matrices. The properties of the operations in these structures and how they are reflected in the number systems of school mathematics should be investigated, especially the use of matrices and matrix operations to record information and to deal with solutions of systems of equations.

Geometry and measurement. Geometry should focus on intuitive, "common sense" investigations of geometric concepts in such a way that general properties emerge and are used as the basis for conjectures and deductions. Later, observations and deductions can be studied more formally as part of a mathematical system. Tessellations, symmetry, congruence, similarity, measurement, trigonometry, and other notions can be investigated through two- and three-dimensional physical models, drawings, and computer graphics, emphasizing visualization. Synthetic, coordinate, and transformational geometry should be revisited with an emphasis on solving problems. The need for assumptions, for more formal arguments, and for formulating, testing, and reformulating conjectures becomes more evident. Taxicab geometry and geometry on the sphere can be used to study alternatives to Euclidean plane geometry. Dimensional analysis can be used to solve more complex problems involving measurement and attendant conversions.

Statistics and probability. Teachers should learn to use key concepts of descriptive statistics, culminating in personal research projects that include experiences in collecting, organizing, analyzing, and interpreting data and in communicating the results of descriptive statistics to others. The concepts of dispersion and central tendency should be represented using techniques from exploratory data analysis. Relationships between two variables should be represented with scatter plots, and visual techniques for approximating a line of best fit through a scatter plot should be introduced as well. Potential misuses of statistics and common misconceptions of probability should be discussed. The power of simulation as a problem-solving technique for making decisions under uncertainty should be a prominent experience. Experiments involving dice, spinners, random numbers, and computer programs should be used to simulate probability and statistics problem situations. Other topics that should be introduced include fair games and expected value, odds, elementary counting techniques, conditional probability, and the use of an area model to represent probability geometrically.

Concepts of calculus. Teachers should acquire conceptual knowledge of the process of differentiation and integration, including examples of applications of these ideas in the sciences and in modeling and solving problems in mathematics. Functions, graphs, and the notion of limits should be explored, starting with concrete problems such as maximizing the volume of a box that can be folded from a rectangular sheet of grid paper. The rate of change of the volume of the box as a function of the height of the box can be investigated in a way that introduces the concepts of differentiation and integration in an appropriate manner for teachers of grade 5 and above. The concepts of limit and infinity should also be explored for their role in the history of the development of calculus and in the study of geometry.

Additional Mathematics for Teachers in Grades 9–12

Teachers of mathematics in grades 9–12 build on the knowledge students have obtained in grades K–8, provide students with broad experiences in the range of applications of mathematics, and help students extend and formalize their thinking and reasoning. With regard to specific content preparation, the mathematical education of teachers at the 9–12 level should include and extend the material described earlier by including the following:

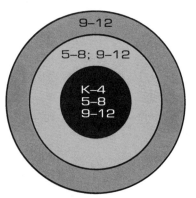

Number systems, number theory, algebra, and linear algebra. Further study of the system of complex numbers should include both geometric (vector) and polar representations of complex numbers and the interpretation of complex solutions to equations. Also, investigations of selected algebraic structures should include groups, rings, integral domains, and fields (including order relations). Topics in number theory should be explored, including modern topics such as coding theory. Because of its wide application, linear algebra should receive extensive treatment. In addition, functions acting on these structures, such as isomorphisms of groups and linear (matrix) functions acting on vector spaces should be investigated.

Geometry. Geometry should be extended to include vector geometry and additional work in synthetic, coordinate, and transformational geometries. Alternatives to the parallel postulate provide opportunities to reveal non-Euclidean geometries. An introduction to the foundations of geometry can provide students with insight into the power of the axiomatic method. The study of geometric transformations, an important manifestation of the function concept, shows the interplay between algebra and geometry. Experiences with linear algebra should be applied to the study of matrix representations of transformations and can shed light on the geometric effects of transformations and the algebraic structure of a set of transformations.

Statistics and probability. For mathematics teachers in grades 9–12, the study of probability and statistics should include both descriptive and inferential statistics and probability from both experimental and theoretical viewpoints. The theoretical probability and statistics should include both discrete and continuous probability distributions and use such distributions to make inferences about populations. On the experimental side, teachers should have extensive experiences using and creating simulations of probability and statistics experiments, both with concrete objects such as dice and spinners and with computer programs. Misuses of statistics and common misconceptions of probability should be discussed. Descriptive statistics should include exploratory data analysis, including the median fit line for a scatter plot, as well as the traditional measures of dispersion and central tendency. Other statistics topics should include confidence intervals, hypothesis testing, correlation, and regression.

Calculus and analysis. Teachers of mathematics in grades 9–12 should have a firm conceptual grasp of the notions of limit, continuity, differentiation, and integration and a thorough background in the techniques and applications of calculus. The development and use of calculus to model and solve problems involving rates of change, optimization, and measurement need to be appreciated as fundamentally important intellectual achievements in the history of mathematics.

Discrete mathematics. The tools and modeling processes of discrete mathematics have gained increased prominence in applications to real-world problems, including those in computer science. Thus it is essential that the mathematical background of secondary school mathematics teachers include attention to symbolic logic, induction and recursion, relations, equivalence relations and functions, and sequences and series. A wide range of modeling applications of graphs and trees should be explored, along with properties of graphs and trees, matrix representations of graphs, and incidence paths in graphs. Other topics should include difference equations and an introduction to combinatorics.

Recommendations for Coursework in Content Mathematics

For teachers of grades K–4, a sufficient understanding of the mathematical topics described (see pages 135–137) cannot be attained with less than nine semester hours of coursework in content mathematics. These mathematics courses assume as prerequisite three years of mathematics for college-intending students or an equivalent preparation.

For teachers of grades 5–8, a sufficient understanding of the mathematical topics described (see pages 135–138) cannot be attained with less than fifteen semester hours of coursework in content mathematics. These mathematics courses assume as prerequisite four years of mathematics for college-intending students or an equivalent preparation.

It is expected that teachers of mathematics in grades 9–12 will have the equivalent of a major in mathematics to gain sufficient understanding of the recommended mathematics (see pages 135–139). It is recommended that experiences showing the variety of applications of mathematics in other disciplines be integrated throughout their study of mathematics. In addition, an emphasis on problem solving and the history of mathematics is essential. The coursework for teachers at this level assumes as prerequisite four years of mathematics for college-intending students or an equivalent preparation.

Since the spirit and content of the coursework described above can be very different from traditional courses, every effort should be made to develop new courses that reflect these differences.

Given the nature of mathematics and the changes being recommended in the teaching of mathematics, teachers at all levels need substantive and comprehensive knowledge of the content and discourse of mathematics. In addition, teachers need to view mathematics through a variety of lenses, including the role and impact of culture, society, and technology and the place of school mathematics within the discipline of mathematics

Vignettes

The instructor has deliberately chosen a problem that will provide opportunities for students to make connections with concepts they have explored in number theory.

2.1 In a university mathematics class for preservice elementary teachers, students have been studying concepts in number theory. The instructor, Dr. Ong, has posed the following locker problem, and students, grouped in teams of four or five, are working on the problem.

> One thousand students have lined up in a very long hall with 1000 closed lockers. One by one the students run through the hall and perform the following ritual: The first student opens every locker. The second student goes to every second locker and closes it. The third student goes to every third locker and changes its state. If it is open, the student closes it; if it is closed, the student opens it. In a similar manner the fourth, fifth, sixth,...students change the state of every fourth, fifth, sixth,...locker. After all 1000 students have passed down the hall, which lockers are open?

Wanda, Craig, Dina, and Mario have been working together for several minutes, as Dr. Ong circulates among the groups.

Students apply ideas and language from their earlier work in number theory. They communicate with each other and draw attention to patterns through the use of a shared mathematical language.

Wanda: So the ninth person goes to locker nine and opens it.

Craig: What about the factors involved?

Dina: Seven is going to change the state of....Seven stayed open until the seventh person got there. Five stayed open.

Wanda: Those are primes.

Mario: So all primes stay open until the person changes the state. So we know that all primes are closed.

Dr. Ong approaches the group.

Students search for patterns as part of their problem-solving strategy.

Craig: One, four, and nine are open.

Dina: These are perfect squares.

Wanda: Let's try four squared.

Craig: Just do sixteen.

Wanda: You can't just do sixteen, because you might have multiples you have to close or open before sixteen.

They see making conjectures as a natural part of their work in this class.

Wanda (turning to Dr. Ong): We're going to conjecture that perfect squares are open. Primes are closed.

The instructor helps students develop a disposition to question, investigate, and justify. He connects students' discoveries and generalizations to further questions.

Dr. Ong: Why? (When the group offers no explanation, Dr. Ong continues.) You have an interesting conjecture, but why? What is so special about square numbers? What is it about the structure of primes that causes lockers with those numbers to be closed? (He moves on.)

Dina: Well, the primes get touched by only that person.

Wanda: But why are the square numbers open?

Dina: Let's look at composite numbers.

Students frame their own mathematical questions and apply other mathematical knowledge as they investigate the problem.

Mario: A composite gets hit for each factor.

Craig: Six is two and three, but four is two and two and nine is three and three. Then why aren't all composites open as well?

The group pursued this problem together for nearly thirty minutes before they had an explanation that satisfied them. Later, when they shared their results with their classmates, they recalled a particular break-through.

Wanda: It took us a long time to consider the importance of all factors of the number.

Mario: Yeah, four has one, two, and four as factors. But six has one, two, three, and six as factors.

Dina: Now it seems so simple. Squares have an odd number of factors and other composites have an even number of factors.

Craig: That repeated factor gave us fits.

Yuko had been working in a group that also had found patterns among the squares, but she wasn't entirely convinced by the discussion so far: "I can see how it works for squares like 4, 9, and 25, but how do you know that a very large square number like 576 would be open?"

Other members of Yuko's group added their own questions:

What if there were more than 1000 lockers?

Even if there were only 1000 lockers, what would be the largest locker number that would be open?

What about the largest number of different people that touch the same locker?

School mathematics is embedded within the larger context of the discipline of mathematics.

Students recognize a need for providing convincing arguments to support conclusions reached through inductive reasoning.

Hearing and discussing other people's ideas and questions stimulates students to generate questions that extend the problem.

2.2 A group of high school mathematics teachers has been meeting twice a month at their school for a seminar with mathematicians and mathematics educators from a nearby university. These teachers have been using computers in their geometry classes for the past year and a half, and the seminar provides them with opportunities to discuss what is happening in their classrooms as they think about new ways of teaching and learning.

The computer software allows students and their teachers to construct geometric shapes and to make measurements of lengths and angles and computations based on these measurements, thus providing an environment for open-ended exploration and discovery of patterns and relationships.

Searching for patterns and relationships is a powerful way for students and teachers alike to extend their knowledge of mathematics.

Although the teachers have been excited about their use of computers in geometry, many have voiced frustration in trying to make decisions about appropriate tasks for their students. Some teachers have been most comfortable focusing student attention on specific relationships, while others are dissatisfied with activities structured to lead students toward a particular "discovery." At times, many teachers have felt their own knowledge of geometry inadequate to deal with questions and conjectures that arise from open-ended explorations.

It is not technology alone that helps teachers and students grow in their understanding of mathematics; how technology is used is critically important.

Gloria described a task she assigned her class early in the year: "I wanted my students to learn that the sum of the angle measures in a triangle is 180 degrees, so I had them construct a lot of triangles on the computer and record the angle measures. The software made it possible to collect a lot of data quickly and make a generalization. I thought my students would remember the relationship better if they discovered it themselves."

Teachers think about the role of technology in the teaching and learning of mathematics and their reasons for using computers in their classes.

Rich talked about the same task: "I was really reluctant to use that activity because it didn't seem like exploration. It made me feel that I would be directing the students toward a single result and not really taking advantage of the technology. But when Gloria told me about some of the things her students came up with, I thought it might lead in some interesting directions. I was amazed at what happened. My students didn't just see what I thought they would see; many of them went off in all sorts of directions exploring other shapes. One even asked about a circle! I wasn't quite sure where to go with *that* question, but it certainly seemed intriguing—and it took us into lots of other ideas when we discussed it in the seminar."

Student questions and conjectures generate new ideas for teachers to pursue with their students and with other teachers. The seminar allows teachers to explore questions with each other and with university faculty in a supportive environment that connects with their classroom teaching.

Constanza remembered a lesson that was especially important to her: "One of my students had constructed a shape on the computer screen that he said looked three-dimensional. We took off on a discussion of geometric models and representations of shapes, something I hadn't really expected to get into during that lesson. As we were talking about two- and three-dimensional shapes, Jan asked about a line. Well, a lot of the students thought that was boring, but then Raoul held up a paper clip and said he thought it was two-dimensional. And another student said that if you bent the paper clip it would be three-dimensional."

Using her knowledge of representations, the teacher takes advantage of a student's observation to discuss a topic that she considers an important part of the school geometry curriculum.

"That set off a bunch of conjectures, with students coming up with good reasons for why the bent paper clip could be considered one-, two-, or three-dimensional. There was a lot more there than I had anticipated, and I thought it would be a great topic for discussion in the seminar. It made us think a lot about representations and how we describe and define geometric shapes."

Teachers, as well as their students, develop their knowledge by framing mathematical questions, conjectures, and arguments and communicating mathematical ideas with others.

The seminar has been a place where teachers can share their struggles with colleagues and university faculty and develop meaningful activities for their students. For many of the teachers, one of the most valuable aspects of the seminar has been the opportunity to extend their own understanding of geometry.

Teachers develop their knowledge of mathematics in many different settings, including school-based programs and their own classrooms as well as in university courses.

2.3 Dr. McPhee teaches a course on the teaching and learning of secondary school mathematics in a university's teacher education program. All his students hold bachelor's degrees and have taken a considerable amount of mathematics, at least the equivalent of a mathematics minor. Many of the students have a major in mathematics.

One focus of this course is the prescribed mathematics curriculum for grades 8–12. The students each have their own copy of the recently revised curriculum guide and are given assignments designed to familiarize them with its organization of topics into strands and scope and sequence.

As they begin their study of the curriculum strand called "Data Analysis," some students are curious about terms such as *stem-and-leaf diagram, box-and-whiskers plot, 90% box plot,* and *median-median line.* In class Dr. McPhee asks the students to indicate their familiarity with these terms. In almost all cases the terms are new to the students, but many think they can figure out the meanings from the context and on the basis of their previous study of statistics.

Dr. McPhee plans class activities that will introduce current conventions of terminology and practice in statistics. Although the focus of the activities is on the mathematics of the curriculum strand being studied, the tasks are taken from resource books written for teachers in which the problem settings are likely to be interesting to secondary school students.

In one of the activities, students are provided with data about several communities' exposure to waterborne pollution from an atomic energy plant that produces plutonium and the number of deaths due to cancer in each community. This plant, which is located about 500 km from the university, has recently been in the local news as questions have been raised about airborne pollution from the plant.

As part of the class, students prepare a scatter plot of the data and carry out the steps of the paper-and-pencil procedure for drawing the "median-median line" to show the relationship between the index of exposure to pollution and the number of cancer deaths. The students also use computer software provided with the resource book to execute this procedure. One student in the class has a hand-held calculator with the regression function built in and uses it to perform a similar analysis based on "least squares."

These students' background in mathematics is generally broad and deep, but they are only beginning their formal study of school mathematics as a subset of mathematics.

Statistics is an area of mathematics in which there have been substantial changes in the nature of the subject and its applications in society.

The prospective teachers' prior study of mathematics has prepared them to be flexible in understanding new developments and applications.

The students see timely and relevant applications of mathematics in society.

The students consider the changes in the ways mathematics is taught resulting from the availability of technology.

Students observe the connections (similarities and differences) between the median-median line, which is new to them, and the least-squares line, with which many of them are familiar.

STANDARD 3:
KNOWING STUDENTS AS LEARNERS
OF MATHEMATICS

The preservice and continuing education of teachers of mathematics should provide multiple perspectives on students as learners of mathematics by developing teachers' knowledge of—

♦ *research on how students learn mathematics;*

♦ *the effects of students' age, abilities, interests, and experience on learning mathematics;*

♦ *the influences of students' linguistic, ethnic, racial, and socioeconomic backgrounds and gender on learning mathematics;*

♦ *ways to affirm and support full participation and continued study of mathematics by all students.*

Elaboration

Learning is an active, dynamic, and continuous process that is both an individual and a social experience. Children are naturally inquisitive and have a desire to learn. Their early experiences reflect the excitement of discovery. In school, however, limitations of time, place, and perceptions often constrain what is natural as children encounter environments that are not responsive to them as learners.

The study of general principles of teaching and learning is insufficient for teachers of mathematics because it does not include consideration of the nature of mathematics and of current research on children's mathematical thinking and its implications for instruction. Children build a variety of perceptions of mathematics as they learn. Some of these perceptions are confused or incomplete; others are remarkably effective. Teachers need opportunities to examine children's thinking about mathematics so that they can select or create tasks that can help children build more valid conceptions of mathematics. Developing multiple perspectives on students as learners of mathematics enables teachers to build an environment in which students may learn mathematics with appropriate support and acceptance.

Professional development programs, both preservice and in-service, should incorporate current theories and research from mathematics education and the behavioral, cognitive, and social sciences as they relate to mathematics learning. For example, central to current theories is the view of the learners as active participants in learning. Learners construct their own meaning by connecting new information and concepts to what they already know, building hierarchies of understanding through the processes of assimilation and accommodation. Mathematics is learned when learners engage in their own invention and impose their own sense of investigation and structure.

The implications of such research and theory building to teaching are continually unfolding as new results from research and practice provide new insights and directions for our understanding. Programs for teachers should help them develop habits of mind that include becoming active researchers in their own classrooms as well as users and interpreters of research as it relates to their everyday teaching. Teachers must be able to interpret research related to instructional issues in order to determine how these issues can be addressed in their teaching. A sampling of current instructional issues includes—

♦ the role of number sense, counting, and manipulative materials;

♦ language and its impact on early mathematics learning;

♦ the implications of children's informal mathematics concepts about size, shape, and space as well as number and chance;

♦ the role of calculators and computers;

♦ probabilistic and proportional reasoning;

♦ the role of variable and function;

♦ the inclusion of discrete mathematics.

These are but a few of the many instructional issues that will continue to be investigated as we strive to improve the teaching and learning of mathematics.

With the help of the available technologies, it is possible to bring the study of children's mathematical thinking "alive" in new ways. Videotapes may be used to portray developmental sequences in learning or to demonstrate assessing students' developmental levels on the basis of specific learning tasks such as those prescribed by learning theorists or designed for new research on students' learning. Indeed, computer-controlled videodisk options make it possible to develop interactive learning environments that teachers may explore to better understand children's thinking in different classroom environments.

In addition, clinical experiences such as interviewing children one-on-one or in groups allows teachers at any level to appreciate what can be learned by talking to students. In conjunction with seminars, courses, or other professional development activities, practicing teachers can learn about current research on children's understanding of mathematics concepts and can validate their knowledge of their own students in order to build deeper understanding of the research and its implications. In follow-up seminars, teachers have opportunities to report and discuss their findings. Changed perceptions about what their students can and cannot do affects teachers' attitudes and beliefs about their students and about their teaching strategies. The importance of teachers' knowledge of how students learn mathematics cannot be minimized. Such knowledge provides direction for the kinds of learning environments that teachers of mathematics create, the tasks they select, and the discourse that they foster.

Teacher expectations have significant impact on what happens to children in school. Teacher expectations are founded on knowledge and beliefs about who their students are and what they can do. Teachers' understanding of the impact of students' age, abilities (both mental and physical), interests, and experience on their learning of mathematics are all important ingredients in building perceptions of students as individuals.

How does mathematics appear to an eight-year-old? A twelve-year-old? A fifteen-year-old? Teachers must simultaneously be able to perceive mathematics through the minds of their students while they perceive the minds of their students through the mathematics in which they are involved. Such a perspective requires a thorough knowledge of children's developmental characteristics that emphasizes children's patterns of intellectual, social, and emotional growth. Beyond a general, comprehensive overview, teachers of the middle grades and secondary levels need a more detailed understanding of adolescence. Such understandings at all levels must be interwoven with teachers' own developing knowledge about how children learn mathematics.

Teachers' beliefs about students often are tied to their perceptions of students' intellectual abilities. Yet the research on ability grouping calls into question current tracking practices. For example, the research indicates that heterogeneous groups make sense in elementary schools. Further, the results of homogeneous grouping in secondary school do not justify its strong support among teachers (Kulik and Kulik 1982; Oakes 1985; Slavin 1986]. Teachers need knowledge about, and experience with, using alternative strategies such as cooperative and team learning that permit them to work well in heterogeneous environments.

Another problem with tracking in mathematics is that students who have difficulty during their elementary and middle school years with traditional paper-and-pencil computation, both arithmetic and algebraic, often are limited in their access to advanced mathematics. However, computational competence is not always a valid measure for success at advanced levels of mathematics. Hypothesizing, approximating, estimating, reasoning, problem solving, and communicating are skills and abilities not often tapped or promoted through traditional computational work. In affirming and encouraging full participation by every student, issues surrounding false scope and sequence barriers that establish inappropriate prerequisites must be considered in professional development activities with teachers.

Teachers also need to understand the importance of context as it relates to students' interest and experience. Instruction should incorporate real-world contexts and children's experiences and, when possible, should use children's language, viewpoints, and culture. Children need to learn how mathematics applies to everyday life and how mathematics relates to other curriculum areas as well.

The ability to recognize and enfold mathematical aspects of ethnic and cultural identity helps in providing an impetus for further study of mathematics. Providing students from underserved and underrepresented groups who lack the observed presence of role models with other means of motivation and incentive for study is one way to do this. References to the contributions made to the discipline by members from underrepresented groups can partially meet this need.

> It is important to note that culture-sensitive does not mean a focus on the traditional arts, foods, and folklore of a group. Instead culture-sensitive means sensitivity to "relatively subtle aspects of interactional etiquette [that] are likely to go unrecognized by non-minority teachers." (Erickson and Mohatt 1982)

Language and its role in students' understanding and doing mathematics needs attention in programs for the development of teachers. Students may lack appropriate vocabulary and syntax to express themselves mathematically but still be able to learn and demonstrate sophisticated knowledge of mathematics. In some circumstances, students' understanding of the language used to communicate mathematics may be incomplete or incorrect; these misunderstandings can create subtle barriers to success in the mathematics classroom. Teachers' knowledge of their students' cultural backgrounds and the implications of this knowledge for their teaching is crucial in recognizing the impact of language on learning. Beyond this, teachers have a responsibility to help students grow in the correct and appropriate use of mathematical language.

Increasing attention has been given to girls' lack of participation in mathematics. Reasons have included a range of hypotheses, such as girls' lack of self-confidence in their mathematical abilities, their association of mathematics with males, and a belief that mathematics is distant from everyday concerns. Earlier considerations of this problem have focused on how to change girls' perception of, and involvement in, mathematics. However, current work indicates that females make sense of information and learn in ways that are significantly different from the traditional approach to teaching mathematics. Programs for mathematics teachers need to provide access to the literature that explicates the problem of engaging girls in the study of mathematics and identifies successful intervention strategies.

> Encouragement and compliments must be specific to be effective, and it is best to praise girls for their ability...not just for their effort. Don't say, "That's okay, you tried." or "Your work is very neat" even though it may be of poor academic quality. (Franklin 1990)

The general issue of discrimination in the classroom is of concern. Such discrimination is often subtle and not intentional, yet it exists. Testing for differential treatment of students is one aspect of addressing the problem of reaching all students. Are there gender, cultural, or racial differences in a teacher's interactions with students in the class? Teachers need help in learning to monitor classroom interactions; a colleague observing or videotaping a class can be of assistance in doing this. Recording instances of positive and negative feedback, disciplinary and social interactions, as well as the name of each student who does and does not receive attention, can provide insights into unconsciously biased behaviors. If inequities are identified, then strategies need to be developed to help a teacher address these concerns. For example, a teacher may decide to keep a list of the names of students in the class and check off those she or he calls on during an instructional period, thus becoming more aware of the participation of all children. Such strategies need to be discussed in professional development activities for teachers.

Grouping of students, classroom climate, choice of materials, topics, activities, testing, and teaching strategies all have impact on how effectively all students consider themselves as involved and active members of the classroom. Each component must be addressed with regard to students' age, abilities, and interests and to their academic, ethnic, racial, cultural, and gender differences. A genuine respect for, and understanding of, students as individuals and as participants in a community of learning is essential to promoting the kinds of experiences that involve all students in mathematics.

Vignettes

3.1 Dr. Williams, the district mathematics coordinator, and several grade 4–8 teachers have formed a mathematics study group that meets monthly. They have been reading about and discussing ways to investigate their students' thinking. In the course of this discussion they began to explore the issue of teaching statistics—particularly the concept of mean. Dr. Williams has brought in a tape of an interview with a sixth-grade student.

Dr. Williams: Before we look at this tape, I want to pose a problem. You should solve the problem and pay particular attention to the strategy you use.

You have nine bags of different kinds of potato chips and you know that the average cost for a bag of chips is $1.38. What might be the actual prices of each of the nine bags of chips?

Dr. Williams has access to a few of the videotapes that have been produced as part of a research project on students' understanding of the concept of average.

By solving the problem themselves, the teachers gain personal awareness of possible strategies before watching a sixth-grade student solve the same problem.

The teachers work individually and then compare strategies with one another. They then spend time discussing their various strategies together. Following this, Dr. Williams plays the videotape, which the teachers discuss in light of their own strategies.

One teacher begins by identifying particular behaviors. In this case, Sara has a real-world context for her thinking. Sara's context for understanding the problem contrasts with another teacher's, possibly reflecting developmental differences.

Mary: I noticed that Sara, the student in the tape, seemed comfortable with the problem. It made some sense to her because she is familiar with the cost of bags of potato chips.

Irv: Yes, right away she said, "Well, I know they can't all be the same price. That's not the way it is in the store." Unlike her, I immediately decided to do a "quick and easy" solution and make each bag the same price—but that certainly isn't real!

Another teacher goes further and begins to consider what Sara actually understands. She realizes that "average" may be a more complex concept than she originally imagined.

Natasha: I was intrigued when she started to figure out actual prices. She knew she wanted some of the prices to be less than $1.38 and some to be more. She also was quite emphatic about stating that there had to be at least one bag with a price of $1.38. Earlier Jody showed us that we could actually have nine bags of potato chips with an average price of $1.38, and none of the bags actually had this price. Sara either wasn't comfortable with this as a possibility, or it may not have even occurred to her. I bet this may be a difficult concept to understand.

The teachers focus on the role of the researcher. How does someone work with a student to gather information about her thinking strategies?

Dr. Williams: What did you notice about the way the interviewer worked with Sara?

Don: Well, he posed the problem and explained the materials that were available, pointing out the pictures of the bags and the markers she could use to record prices. Then he became an observer. Occasionally he asked a question for clarification, but he really didn't involve himself in her work. Gee, it would be difficult for me to know when to intervene and when to keep silent. I would want to jump in with suggestions!

The teachers have an opportunity to work with students, using research tasks from the project considered in the study group.

These teachers are engaged in experiences, such as considering the effects of age on student learning, that can affect their knowledge of research and their dispositions to listen to children and use the information gained to make instructional decisions.

The teachers and Dr. Williams continue the discussion, commenting on the student's response to the problem. Finally, Dr. Williams provides a synthesis from this research project, detailing other problems used in the research and providing an overview of the results of the research that focuses on the developmental differences that were found between the fourth-, sixth-, and eighth-grade students who were interviewed.

The teachers agree to pose the same problem to some of their students and bring the results to the next study group meeting.

Collegial support is an essential part of professional development.

3.2 Darren Hensh and Cora Horatio have been long-time colleagues and often consult each other about their teaching concerns. In a recent discussion Darren brought up a situation that arose when teaching mathematics to his third graders (Ball 1990).

The teacher looks for ways to engage all students and selects a task that will push his students' thinking further.

Darren: Up to this point the work my class has done with division has included dividing things into groups of equal size and talking about leftovers. I felt they were ready to consider fractions by dividing the leftovers up, too. So I posed this problem:

> You have a dozen cookies, and you want to share them with the other people in your family. If you want to share them all equally, how many cookies will each person in your family get?

The teacher knows his students and can use this knowledge to plan activities that are mathematically interesting and enhance the discourse in the classroom.

Cora: Hey, that sounds like a clever problem! Your divisor would vary nicely among the students, providing for a range of interesting solutions, some simpler than others. I often use family size as a context for problems I do with my sixth graders.

Darren: You do? You know, I like the diversity it provides, but I am questioning my choice because I didn't anticipate the interactions it stimulated. As I listened to their discussions, I realized that who to count in a family is not cut and dried, particularly for young children. For example, Enrico wondered if he should be counting the baby that is due soon. Angela included her mother's boyfriend as one of her family members, which confused her friend Jill. And...oh, yes,...Micha's parents are divorced, and his father lives in another state. But he included him in his count. His two partners argued with him, saying that he really couldn't count him if he didn't live with him.

All the students found solutions that were satisfactory, but somehow, I felt uneasy about the nature of their arguments. They seemed intrusive and possibly made some of the students uncomfortable. I am doubting the wisdom of choosing such a problem.

Cora: I know what you mean. They all have very different conceptions of family, so it can be awkward. But I don't think you should avoid issues like this. Not only is family size a realistic and engaging context for such problems, but it is important for children to become aware of the different ways people define family and to respect these differences.

Darren: I agree. I am pleased to have the diverse range of views in the class. It was their tendency to impose their own conceptions on one another that I found troubling.

Cora: When I've used family size with students we begin with a class discussion about who to count. Yes, it is a sensitive issue, but I have found that the children naturally respect one another's lifestyles when approached as part of "finding a definition" that works for all of us. It's very interesting to listen to their own conceptions of family. I always learn so much about each student. Beginning with a discussion like that should help them recognize that there are many ways to think about the family unit.

The teacher is sensitive to children's different family backgrounds, recognizing that the appropriateness of the context of a problem is an important factor to consider when designing and selecting problems.

To encourage full participation, the teacher is trying to build an environment in which all students feel respected and valued. His colleague suggests how he can turn the different family structures into a positive mathematical and social experience for the children.

3.3 The student teachers in Dr. Dreyfus's seminar group are discussing a teaching dilemma described in an article on how teachers manage to teach (Lampert 1985):

"The children in my classroom seem to be allergic to their peers of the opposite sex. Girls rarely choose to be anywhere near a boy, and the boys actively reject the girls whenever possible. This has meant that the boys sit together at the table near one of the blackboards and the girls at the table near the other."

"The fifth-grade boys are particularly enthusiastic and boisterous. They engage in discussions of math problems with the same intensity they bring to football. They are talented and work productively under close supervision, but if left to their own devices, their behavior deteriorates and they bully one another, tell loud and silly jokes, and fool around with the math materials. Without making an obvious response to their misbehavior, I developed a habit of routinely curtailing these distractions from the lesson by teaching at the blackboard on the boys' end of the classroom. This enabled me to address the problem of maintaining classroom order by my physical presence. But my presence near the boys had inadvertently put the girls in "the back" of the room. One of the more outspoken girls impatiently pointed out that she had been trying to get my attention and thought I was ignoring her. She made me aware that my problem-solving strategy, devised to keep the boys' attention, has caused another, quite different problem. The boys could see and hear

The university instructor uses current literature on real teaching situations to help student teachers consider gender equity.

The behavior of fifth graders in male-female interactions is explored.

The teacher recognizes a problem that results in an inequity in treatment in the classroom.

Boys are receiving more attention and hence more help as a result of the teachers behavior management strategy.

more easily than the girls, and I noticed their questions more readily. Now what was to be done?"

The complexity of teaching is confronted in the case.

"I felt that I faced a forced choice between equally undesirable alternatives. If I continued to use the blackboard near the boys, I might be less aware of, and less encouraging toward, the more well behaved girls. Yet, if I switched my position to the blackboard on the girls' side of the room, I would be less able to help the boys focus on their work. Whether I chose to promote classroom order or equal opportunity, it seems that either the boys or the girls would miss something I wanted them to learn."

No algorithms for such decisions exist. The teacher is shown struggling with a real teaching dilemma.

The instructor encourages discussion of alternative strategies.

Dr. Dreyfus opens the discussion by commenting, "Dilemmas are an inherent part of teaching. We may "know" what is desirable in terms of theory, research, and expected practice, but often the route we take to achieve one goal is at odds with another. What are some of this teacher's alternatives?"

The field component of the preservice program provides opportunities for prospective teachers to identify and confront similar issues.

Ian immediately responds, "You know, I have a similar situation in my seventh-grade class. The girls are generally more attentive and less distractible, so I spend more time focused on the boys. I tried mixing up the boys and girls, assigning seats that distributed them around the room. Now they spend more time talking with each other than paying attention to mathematics. I don't feel I have solved the problem."

Another student teacher sees the parallel with her dilemma in handling students with language differences. Possible strategies to help with the problem are shared.

"It's hard to be attentive to all the needs of my students," notes Kadisha, who is working in a sixth-grade ESL class. "I'm particularly aware of the students who have language difficulties. I give them a lot more attention than other students. Maybe the teacher in the article should try some small-group activities. Then she could walk around the room and spread out her attention."

Issues of gender equity must be addressed light of other pedagogical concerns, such as the role of discourse . These students are gaining a sense of the complexity of teaching and of ways in which they need to monitor their own behaviors in the classroom.

"But whole-class discussions are a central part of my mathematics classes," Maureen responds. "By the time the students are in ninth grade the pattern has been established. The boys dominate the discussion and many girls are reluctant to contribute. I find myself asking the girls less challenging questions or changing the tone of my voice in order to encourage them to participate. So in a sense I'm countering one imbalanced situation with another. But I'm hoping to gradually change that."

Student teachers are developing sensitivity to all students in their classrooms and seeking ways to manage the ongoing dilemmas of teaching.

"I don't think we can expect changes overnight," Rika reminds her peers.

As we leave the seminar, other students are sharing strategies for the teacher's dilemma and raising questions about similar issues from their student teaching.

STANDARD 4:
KNOWING MATHEMATICAL PEDAGOGY

The preservice and continuing education of teachers of mathematics should develop teachers' knowledge of and ability to use and evaluate—

♦ *instructional materials and resources, including technology;*

♦ *ways to represent mathematics concepts and procedures;*

♦ *instructional strategies and classroom organizational models;*

♦ *ways to promote discourse and foster a sense of mathematical community;*

♦ *means for assessing student understanding of mathematics.*

Elaboration

> In mathematics the reflective process, wherein a construct becomes the object of scrutiny itself, is essential. This is not because, as so many people claim, mathematics is removed from everyday experience. It is because mathematics is not built from sensory data but from human activity (mathematics is a language of human action): counting, folding, ordering, comparing, etc. As a result, to create such a language we must reflect on that activity, learning to carry it out in our imaginations and to name and represent it in symbols and images. (Confrey 1990, p. 107)

Mathematics pedagogy focuses on the ways in which teachers help their students come to understand and be able to do and use mathematics. This standard identifies several components of pedagogy that are essential to quality teaching. These components act as a series of lenses through which teachers filter their knowledge of mathematics and of students in order to enrich and enhance their teaching of mathematics.

Teachers are responsible for posing worthwhile mathematical tasks. They may choose already developed tasks or may develop their own tasks to focus students' mathematical learning. To do so, they often rely on a variety of instructional materials and resources, including problem booklets, concrete materials, textbooks, computer software, calculators, and so on. Teachers need a well-developed framework for identifying and assessing instructional materials and technological tools, and for learning to use these resources effectively in their instruction.

Such a framework is built from teachers' own understanding of mathematics and what constitutes worthwhile mathematical tasks as well as their knowledge of ways to represent mathematical ideas. Modeling mathematical ideas through the use of representations (concrete, visual, graphical, symbolic) is central to the teaching of mathematics. Teachers need a rich, deep knowledge of the variety of ways mathematical concepts and procedures may be modeled, understanding both the mathematical and developmental advantages and disadvantages in making selections among the various models. In addition, teachers need to be able translate within and between modes of representations in order to make mathematical ideas meaningful for students (Heid 1988).

Representations serve as vehicles for examining mathematical ideas. Not only do teachers need to be familiar with a variety of representations, they must be comfortable with helping students construct their own representations. Designing instruction involves a variety of decisions about the role and use of representations: Should structured representational materials be used? If so, which representational materials provide the most appropriate model for helping develop the concept at this point in instruction? If not, how can students refine the existing models they have been using or develop new models for themselves? Of the various options, what representations are most familiar to students and, therefore, will make sense to them? Numerous other questions surface before instructional choices are complete. Choosing, modifying, or constructing representations are central pedagogical considerations that must be addressed continually.

> Other capabilities suggest more meaningful learning. Can the child...illustrate the rule with physical objects? ...give one or more reasons why the rule is true? ...set up a pattern (using objects or using numbers) through which the rule can be discovered? [These] capabilities go beyond computation; they involve connections between numerical symbols and non-numerical domains, and they make explicit reference to reasoning processes as well as products. (Goldin 1990, p. 46)

Mathematical instruction often is approached in terms of stating and exemplifying rules—the "tell, show, and do" model. Based on the assumption that information can be presented by *telling* and that understanding will result from *being told*, such an approach does not work because it frequently overlooks two crucial developmental components: the process of assimilation and the issue of readiness. Essentially, in this approach, students are "ready" intellectually when the teacher is ready for them to receive the information. *Learning* through such an approach often fails to promote a lack of transfer of mathematical information to new situations.

Teachers need to employ alternative forms of instruction that permit students to build their repertoire of mathematical knowledge and their abilities for posing, constructing, exploring, solving, and justifying mathematical problems and concepts. Promising models for such instruction are all highly interactive. In such models, teachers both model and elicit mathematical discourse by asking questions, following leads, and conjecturing rather than presenting faultless products (Ball 1990; Noddings 1990).

Teachers need to focus on creating learning environments that encourage students' questions and deliberations—environments in which the students and teacher are engaged with one another's thinking and function as members of a mathematical community. In such a community, the teacher-student and student-student interaction provides teachers with opportunities for diagnosis and guidance and for modeling mathematical thinking, while, at the same time, it provides students with opportunities to challenge and defend their constructions.

Teachers need to employ strategies that will help them develop the participation essential to engaging students in mathematics. Increasing the amount of time students spend working together supports the development of discourse and community. Working in groups, students gradually internalize the discourse that occurs, challenging themselves by asking for reasons and, in general, accounting for their own mental work. Another practice that supports students' participation involves shifting responsibility from teacher to student for control of learning by expecting students to make commitments to their answers. Further, students' reflective processes can be developed by focusing their efforts on interpreting problems, describing strategies for solutions, and justifying and defending the results.

Teachers' willingness to be flexible and curious about mathematics with their students is central to their ability to promote mathematical discourse. Engaging in personal discourse with other colleagues about mathematics and mathematics instruction and establishing a classroom environment that encourages engagement in discourse helps teachers deepen, extend, and enhance their knowledge of mathematics and of their students' knowledge of mathematics. Teachers need to experience and reflect on discourse and their own efforts to promote discourse in order to identify what works or does not work, enriching and extending their capabilities to involve their students in mathematical ideas.

> As we need standards for curricula, so we need standards for assessment. We must ensure that tests measure what is of value, not just what is easy to test. If we want students to investigate, explore, and discover, assessment must not measure just mimicry mathematics. By confusing means and ends, by making testing more important than learning, present practice holds today's students hostage to yesterday's mistakes. [National Research Council 1989, p. 70]

Assessment should be an integral part of mathematics teaching. Through assessment, teachers learn how students think about mathematics and what they are able to accomplish. Moreover, students obtain feedback in order to make adjustments and deepen their understandings of mathematics (Stenmark 1989).

Assessment should focus on addressing students' development of mathematical power: their understanding of mathematical concepts and procedures and the relationships between them, and their abilities to reason mathematically and apply their knowledge to a variety of problem situations. Teachers need to align assessment with instructional goals and to consider their purposes in assessment as they select or develop the means of assessment. In addition, teachers need to understand the issues surrounding assessment in general, the arguments related to these issues, the distinctions between classroom assessment and accountability testing, and proposed alternatives for unifying instruction and assessment.

> Every teacher is continually offered a wealth of assessment information during the process of instruction. Many act on this information but few document it. It is through our documented assessment that we communicate most clearly to students which behaviors and learning outcomes we value. [Clarke 1988, p. 19]

The results of assessment—formal and informal—need to be used and communicated. The communication may involve only the teacher using information collected to provide direction for working with an individual student, group of students, or class of students in continuing instruction. Teachers also need to communicate with one another about learning and teaching. Results from assessment often provide the catalyst needed for jointly diagnosing students' understandings and misunderstandings, designing curriculum, planning instruction, or initiating further assessment efforts. Finally, teachers need to evaluate students and communicate this information to students, parents, and others in a school district to provide feedback of a more formal kind that indicates students' understanding of mathematics.

Assessment has a central role in effective teaching of mathematics. Too often, teachers' experiences with methods of assessment are limited to the more traditional "testing and measurement" strategies provided through a preservice course. Given the growing awareness and efforts for change, there is a strong need to integrate the understanding and use of alternative methods of assessment as an ongoing topic throughout teachers' educational life.

The aspects identified in this standard as "mathematics pedagogy" are integral to the effective teaching of mathematics. Teachers' knowledge and their ability to use and evaluate these components develop over time. Decisions about instructional materials are intimately associated with decisions about ways to represent mathematics concepts and procedures. Choices for instructional strategies and classroom organizational models both evolve from and influence such decisions. Finally, the discourse of the classroom and the need for ongoing assessment also are part of this process of dynamic interaction that results in knowing mathematical pedagogy.

Vignettes

4.1 As part of the state's efforts to align school assessment with NCTM's *Curriculum and Evaluation Standards for School Mathematics*, a two-day conference is being held. Many schools have sent teacher representatives who will, in turn, sit on district committees to design K–12 assessment programs. The conference comprises a number of minisessions, each demonstrating an alternative form of assessing what children know and understand about mathematics.

We first visit the session on interviewing.

Teachers are being asked to think very differently about both the purposes of, and ways to conduct, assessment. In reality, many teachers will be documenting what they may do informally now.

Presenter: Assessment is often thought of as being synonymous with paper-and-pencil testing. We would like to broaden this conception to include a multitude of ways of determining what a child actually understands.

The presenter first shows the audience a subtraction paper completed by Myra, a first-grade student. There are twenty-five problems dealing with subtracting a single digit number from a number in the range 5-19.

Myra has four problems circled that have incorrect answers. The audience is asked to hypothesize about what might be Myra's difficulties. After some discussion, it seems clear that the errors are inconsistent, and most agree that probably Myra just got bored.

Next, they view a videotape of a three-minute interview with Myra in her classroom. In the tape, Myra is presented with a large number of cubes (in this case 36) and asked how many there are. We see Myra counting as she points to individual cubes:

Myra: 1, 2, 3, 4, ..., 26, 27, 28, 97, 96, 95, 98, 99. Yes, there are 99 cubes.

Interviewer: Myra, you remember how we grouped cubes by making sets of ten. (Myra nods enthusiastically.) Do you think you could do that for me right now?

Myra begins to put cubes together. The tape zooms in as Myra is counting the remaining single cubes, having made three sets of ten cubes.

Myra: 1...2...3...4...5...6. Six. (Myra looks up and smiles triumphantly).

Interviewer: That's very nice. Now, can you tell me how many you have all together?

Myra wrinkles her forehead in concentration.

Myra: (Smiling). There are nine all together.

Some of the teachers in the audience react to the tape.

Lauren: Look! She counted the six single cubes and the three sets of ten cubes, which results in her answer of nine. I don't think she has a sense of the structure of base-ten place value.

Ed: Yes, the fact that she counted in this way, even after she had actually put them together herself, makes me think she doesn't understand grouping.

Anita: And from looking at her paper we thought she was doing fine. Taking the time to talk and ask students to explain their answers and to watch them work in this form of mini-interview is essential to knowing where students are. It is a technique that can be used at all levels. I would like the teachers at my school to see this video and learn some ways to interview their students.

Next we visit a session on performance-based assessment:

With the help of several high school students in grades 10–12, the presenter engages the teachers in a performance-based assessment activity. In small groups the teachers gather around two or three students and watch as they work on the following problem:

> You are given a square that is 5 units on a side. Your task is to draw another square inside this square that is half its area. Write an explanation of how you know that your new square is half the area of the original square.

Afterwards the teachers are free to ask the students questions about their strategies and solutions. This is followed by a discussion during which the teachers share and compare what they feel they learned from watching the students work.

If paper-and-pencil tests are the only strategy for gathering information, teachers may know very little about their students' understandings of mathematics.

A more open-ended task allows the student flexibility in interpreting the task and demonstrating her understanding, and the teacher can quickly identify information for instructional planning.

Note: Myra counted to 28. Then she jumped to 97, followed by 96.

The interviewer deliberately maintains a supportive and encouraging attitude. The goal is to find out what the child understands and not to intervene.

Quick interviews can provide a degree of knowledge about students' understanding that cannot be provided through paper-and-pencil tests.

Employing alternative assessment strategies takes time. Collegial interactions and professional development programs can help in making changes.

Performance-based assessment involves students, working individually or in groups, solving a problem that may take 15–30 minutes up to a few days.

Such problems are open-ended, permitting students to explore a variety of options. Without a prescribed "rule" or limited expectations concerning results or solution strategy, teachers can find out quite a bit about how their students think.

Marc: This really is a good problem for a variety of students. You could do it with little "formal" mathematical knowledge.

Allison: It also was interesting to see the students' false starts and the difficulties they encountered. Listening to the students' explanations was particularly telling about what they did and didn't understand. It's so easy to accept an answer that looks good without investigating further.

Lu: I really need to think about how I would use this to assess students' knowledge. I could consider what we know about students' mathematical and problem-solving abilities based on their solutions. Clearly, using just one problem is not the answer. I do need to find more time to think about performance-based assessment in greater detail.

Teachers need to think about the purposes of assessment and what information a particular task adds to their understanding of students' knowledge and problem-solving abilities.

4.2 Fred is a first-year teacher. His preservice preparation has successfully convinced him that he needs to create a problem-solving focus in his teaching. Yet he is not prepared for the ways his students react to his efforts to involve them (Brown, Cooney, and Jones 1985).

In a series of lessons, he has students experiment with dice in order to help them appreciate probability and, eventually, insurance and mortality tables. Both the topic and the methods are not successful, and we hear him discussing his difficulties with a mentor mathematics teacher, Jim, who has been assigned as part of a special program in the school district directed at helping new teachers.

Fred has begun without assessing his students' understanding and conceptions of mathematics.

"My students don't view probability as *real* mathematics. They consider the activities with dice as playing games and think we are wasting time. It is really ironic, because if I was doing the types of things they wanted to do, they would be turning around in their seats and talking. So it's a no-win situation."

Jim, a teacher who is known for his ability to employ a variety of strategies to engage his students in mathematics, is sympathetic. He was a first-year teacher himself. And he certainly does not want to discourage Fred's desires to go beyond the textbook to engage his students' interest.

Fred feels a need to meet his students "on their level." However, the students don't feel as though Fred is addressing their need to learn "real mathematics." The environment in the class is not working well.

Fred continues, "My professors in college modeled and expected us to think about and use a variety of problem-solving techniques to engage students. When they were around—and my supervising teacher was around—I felt a little rough around the edges, but I didn't expect this to happen. Somehow, we never talked about this kind of student reaction. How can I tackle something that they consider "real mathematics" and do it in a problem-solving way?"

Fred is experiencing the first-year syndrome of "flying solo" without the ready availability and commitment of several professionals to assist him. He also views traditional textbook mathematics as not applicable to problem solving.

Jim offers an observation. "It sounds like you may be thinking about problem solving and what your students call 'real mathematics' as two distinctly different ways to think about mathematics instruction. It is possible to rephrase questions to encourage more open-ended problem investigations even with the more traditional topics in mathematics."

The mentor teacher identifies the two extremes of Fred's understanding of what constitutes successful ways to promote students' involvement.

Jim opens the textbook to a page that discusses finding the perimeters of a number of different figures. In each case, the figure is shown and each side of the figure is labeled with its respective measure.

Jim notes, "This is pretty straight-forward. But what if you posed some different questions? How about this one: Draw a rectangle with an area of twenty square units. What's the smallest perimeter you can have with a rectangle having an area of twenty square units? What's the largest?

Alternatively, you can have your students construct rectangles with perimeters of twenty units. Now, what's the largest area possible? The smallest area possible? Going further, what happens if we consider other values for the areas or perimeters? What theories can the students build?"

Fred is silent for a short while. "Well, that is a good strategy. But I really could use some help looking at the textbook and thinking about integrating experiences with problem solving."

"Let's plan to meet during our joint planning session tomorrow," Jim responds. "We'll begin to map out what content you want to consider during the next few weeks. Together we'll explore different instructional strategies that may help you engage your students in more problem solving even while they are doing what they call 'real mathematics.' As part of our discussion, let's also talk about how you can do some informal assessing of your students' learning during this process."

The mentor teacher helps Fred consider ways to engage students in problem solving while he includes more traditional content. Question posing is an essential element in building classroom discourse. Jim's instructional strategies provide Fred with a window to make some changes in what and how he teaches.

The mentor teacher intends to help Fred use his instruction as a vehicle for assessing students understanding so that he can find ways to modify and adapt his teaching more appropriately.

4.3 In his course for preservice teachers on the teaching and learning of secondary school mathematics, Dr. Glendon emphasizes mathematical problem solving. He includes opportunities for his students to solve nonroutine problems, to discuss problem-solving strategies, to consider ways of helping students become better problem solvers, and to assess secondary school students' problem-solving performance.

The importance of mathematics as problem solving in the Curriculum and Evaluation Standards is reflected in this course on mathematical pedagogy.

As a culminating activity the university students have been assigned the task of choosing a rich problem and using it in an interview with a student or pair of students during one of their pre-practicum visits to a secondary school. They are to prepare a written report on the qualities of the students' understanding and problem solving.

This activity focuses on student assessment.

Three of the university students have chosen to interview tenth-grade students using the following problem, which they found in one of the resources provided by Dr. Glendon.

> A textbook is opened at random. The product of the numbers of the facing pages is 3192. To what pages is the book opened?

The three students, Michelle Tremblay, Peter Marshall, and Ruth Wong, have decided to make a joint report on their findings and have chosen to organize their presentation using Polya's description of the problem solving process, beginning with understanding the problem.

This problem has the potential to shed light on secondary school students' understanding of several different mathematical ideas, their problem solving processes, and their ability to make connections.

Michelle reports that the student she interviewed at first seemed confused by the problem and reluctant or unable to get started with it. So she asked, "Can you explain in your own words what the problem is telling you?" and later, "What operation was done with the two page numbers to get the answer 3192?" Michelle comments, "The questions I asked weren't really hints; they were just the encouragement the students needed to get into it."

In this activity, the students consider questioning techniques. These techniques, being developed in the context of evaluation, can also be applied in teaching.

Peter begins the discussion of students' strategies for solving the problem by describing the guess-and-test approach taken by the pair of students he interviewed. "Darlene always guessed two numbers that were consecutive (like 81 and 82, 46 and 47) but Jill made wild guesses including pairs of numbers that were not even consecutive (like 31 and 100, 62 and 75). After Darlene and Jill had each worked separately for a couple of minutes, I asked them to compare their results so far," Peter reports. "I didn't even have to tell them what kind of guesses were better—right away Jill realized that her nonconsecutive pairs weren't

The prospective teachers have observed the students' strategies for solving the problem and have also distinguished between students who use a strategy well and those who use it inefficiently or inappropriately.

helpful, and she was able to suggest the next guess—and that turned out to be very close."

The difficulties that students encounter may reveal conceptual difficulties, misconceptions, or common error patterns that teachers can anticipate in their teaching.

Michelle reports that about half the students interviewed used an algebraic approach; some did so on their own initiative, while others did so following a general hint. "They didn't have any trouble deciding to let x be one page number and $(x + 1)$ be the other, and they all got $x(x + 1) = 3192$," she says. "But from there, different students had different problems. Beth wrote

$$x(x + 1) = 3192$$
$$(x + 1) = \frac{3192}{x}$$

but didn't know how to go on from there.

Bruce wrote

$$x^2 + x = 3192$$
$$x^2 + x - 3192 = 0$$
$$(x + \underline{})(x - \underline{}) = 0$$

The interview also provides information about the students' persistence when they encounter problems that are structurally similar but numerically more difficult than typical ones.

but got stuck when he tried to factor the quadratic. I think he just gave up because 3192 was such a large number. The coefficients in the quadratics he had seen in class were always much smaller numbers."

The prospective teachers have identified another approach to this problem. This approach represents a mathematical connection that they as teachers may want their students to explore.

Peter considers the idea of developing a plan. "Michelle, Ruth, and I thought that students might use a factorization approach in solving this problem, but none of them did. We expected that the students might try to break down 3192 into its prime factors and then reassemble the pieces to make two factors that are consecutive numbers."

Selected follow-up questions help evaluate what Polya has called "looking back." Another way of thinking about these questions is to see them as helping students to develop mathematical connections.

Ruth notes that most of the tenth-grade students interviewed by the three university students used calculators as they tried to solve the problem by a guess-and-test approach, which her student, Marc, referred to as "the long way." However, none of them thought of the square root operation. "I was a little disappointed that no one realized that $\sqrt{3192}$ would be a good estimate for the page numbers," she says. "As a follow-up question, after Marc had found the correct page numbers by trial-and-error, I asked, 'If you had known the value of the square root of 3192, would that have helped you to solve the problem?' But he didn't seem to see the connection I was hinting at."

Cross-Referencing Standard 4 with Vignettes in Sections 1 and 2

The vignettes presented in the first section, "Standards for Teaching Mathematics," reflect teaching episodes documenting the four components of tasks, discourse, environment, and analysis. As might be expected, these vignettes demonstrate teachers' applications of the components detailed in this standard as well. The chart below provides a cross-reference from the subtopics of this standard to the vignettes presented in the first section. For example, vignette 1.3 provides an example of using the calculator; as part of vignette 2.3, a teacher evaluates a mathematics textbook, making decisions about ways to integrate its use into her instructional plan. Both these vignettes demonstrate teachers using and evaluating instructional materials and resources.

As part of the preservice and continuing education of teachers of mathematics, the vignettes charted below may be helpful as part of teachers' efforts to develop the "knowledge of, and ability to use and evaluate" instructional materials and resources, ways to represent mathematics concepts and procedures, and so on.

Standards for Teaching Mathematics—Vignettes

	Vignette number and page number																		
	1.1 28	1.2 29	1.3 30	2.1 36	2.2 38	2.3 40	2.4 42	3.1 45	3.2 47	3.3 49	4.1 52	4.2 53	5.1 58	5.2 59	5.3 60	6.1 64	6.2 65	6.3 66	6.4 66
◆ Instructional materials and resources, including technology		•	•			•	•		•			•							
◆ Ways to represent mathematics concepts and procedures		•		•						•	•	•	•						
◆ Instructional strategies and classroom organizational models	•				•	•							•	•	•				•
◆ Ways to promote discourse and foster a sense of mathematical community		•	•	•	•	•	•	•	•	•	•	•	•	•	•				
◆ Means for assessing student understanding of mathematics							•										•	•	

In addition to the vignettes from section 1, the vignettes for Standard 8 (8.1 and 8.2) in section 2, "Standards for the Evaluation of the Teaching of Mathematics," provide further examples of ways to represent mathematical concepts and procedures.

STANDARD 5:
DEVELOPING AS A TEACHER OF MATHEMATICS

The preservice and continuing education of teachers of mathematics should provide them with opportunities to—

♦ *examine and revise their assumptions about the nature of mathematics, how it should be taught, and how students learn mathematics;*

♦ *observe and analyze a range of approaches to mathematics teaching and learning, focusing on the tasks, discourse, environment, and assessment;*

♦ *work with a diverse range of students individually, in small groups, and in large class settings with guidance from and in collaboration with mathematics education professionals;*

♦ *analyze and evaluate the appropriateness and effectiveness of their teaching;*

♦ *develop dispositions toward teaching mathematics.*

Elaboration

> I recently saw a young man in a video store; he reminded me that he had been in my class during my second year of teaching third grade. He was telling me that he is now a teacher. I told him that I had learned so much since that time; that I regretted my lack of experience when he was in my class. He really surprised me. He said that he had enjoyed my class even then because I loved math and that he had not had another teacher that truly loved math until he took trigonometry in high school. (A middle-grades teacher)

This standard addresses issues that are at the heart of teaching. The goal of teacher education is to "light the path" for those who follow, providing directions on how to plan and teach mathematics. It is the practice of teaching, the growing sense of self as teacher, and the continual inquisitiveness about new and better ways to teach and learn that serve teachers in their quest to understand and change the practice of teaching.

The nature and kinds of teaching experiences that should be part of the preservice and continuing education of teachers of mathematics are varied and numerous. For teacher candidates, this involves opportunities to work one-on-one or with small groups of students in clinical settings that permit them to focus on interviewing or microteaching. They need a sequenced program that provides them with opportunities to be in classroom settings for a variety of purposes and with increasing levels of responsibility. Finally, they need long-term placements that permit them to become the teachers of students under the guidance and support of both a cooperating practitioner and a mathematics educator.

During the first few years, teaching is an intensely focused experience that centers on the students for whom the teacher is responsible and on the teacher's growing sense of self as a teacher of mathematics. Colleagues and supervisors can function informally and formally as resources during this time of transition between the structured and guided preparation to teach and the comfort provided by a few years' successful experience with teaching. Indeed, beginning teachers often welcome and seek the advice of more experienced teachers to give guidance and provide some diversity in models of how to teach.

Experienced teachers have different needs. They have a general frame that surrounds their picture of teaching and understand the ebb and flow of the learning process as it proceeds daily, weekly, and monthly throughout the school year. They are better able to anticipate timing, overall organization and management, and student response. Their repertoire of instructional methods has "filled out," and they often can successfully anticipate what works and does not work in the classroom. Nevertheless, they may find times and opportunities when they turn to colleagues and supervisors to assist them in assessing their teaching and making changes. In addition, when teaching new material or trying out new methods of teaching, teachers are in a position for self-evaluation regarding what works and does not work for them.

Good mathematics teaching is enhanced by conversations with colleagues and supervisors who know mathematics and have been successful in teaching mathematics. Preservice teachers should have opportunities to teach with exemplary mathematics teachers. They should be supervised by teacher education faculty who know mathematics and are experienced mathematics teachers themselves. Practicing teachers also should involve colleagues or teacher educators with backgrounds in mathematics teaching when they are exploring new ways to teach or seeking feedback on current teaching strategies. Mathematics has its own content and pedagogy. Only those knowledgeable about the associated special issues and experienced in the field should serve as mentors or supervise teachers' clinical and field-based learning experiences.

Essentially, being a teacher of mathematics means developing a sense of self as such a teacher. Such an identity grows over time. It is built from many different experiences with teaching and learning. Further, it is reinforced by feedback from students that indicates they are learning mathematics, from colleagues who demonstrate professional respect and acceptance, and from a variety of external sources that demonstrate recognition of teaching as a valued profession. Confident teachers of mathematics exhibit flexibility and comfort with mathematical knowledge and commitment to their own professional development within the larger community of mathematics educators.

Vignettes

5.1 In her mathematics methods class, Dr. Palmer has been trying to help prospective elementary teachers learn to "listen mathematically" to children. They have been reading case studies of young children, watching videotapes, and reading theoretical pieces on how children learn mathematics. This week Dr. Palmer assigned students the task of examining closely some aspect of students' understanding in their field classrooms. This afternoon she is meeting with Mr. Konook and five prospective teachers who work in his fourth-grade class to discuss their observations.

Observing and interviewing children can help teachers revise their assumptions about how students learn mathematics and learn to interpret students' words, representations, and ways of putting things.

The university faculty and cooperating teacher work together to help preservice teachers develop as teachers of mathematics.

Damon immediately brings up a conversation he had with one of the students that afternoon. "The last question on the board was to show which was more, ⅖ or ⅓. Tia wrote

When I asked her to explain, she drew this picture:

The student teacher is learning aspects of informal assessment of children's thinking.

The classroom observations show the variety of strategies that children employ to make sense of fractions. Through observations, questioning, and listening to students' explanations, the prospective teachers uncover the thinking underlying their approaches. Going beyond their written answers, they learn to probe the depth of students' understanding.

Because she didn't draw each piece the same size, her picture of ⅖ was indeed larger than ⅓."

"That's interesting," Lisa said. "Latalya got the same answer but for a different reason. She drew this picture:

dogs   hamsters

and said that ⅖ is more than ⅓ because one-third is the same as one out of three and two-sixths is the same as two out of six. We have three dogs, one of them is black. We have six hamsters, two of them are black. We have more black hamsters than black dogs."

"Wait," Maura said, "I don't understand. One of three dogs and two of six hamsters are both one-third."

Listening to children raises questions about the mathematics itself. This provides an opportunity to develop the preservice teachers' knowledge of fractions.

"But she is right," Peter exclaimed, "because she does have more black hamsters than black dogs."

Lisa explained, "I wouldn't have understood how Latalya got her answer if I hadn't asked. I just assumed that she didn't understand the problem. If that were a question on a test, her answer would not reflect what she does know about fractions."

Opportunities to examine student thinking encourage prospective teachers to assess the advantages and disadvantages of various forms of assessment.

"But we know that ⅓ *is* equivalent to ⅖ when we are talking about two equal-sized things. Don't we want them to see that? Isn't this just confusing them? " asked Maura.

The discussion raises teaching issues that focus on how children learn and confront the prospective teachers with subtleties of the mathematics.

Kamisha added, "If Latalya has twenty cats and two of them are black, should she say that ²⁄₂₀ was the same amount as ⅖?"

"What do you think she would do?" Dr. Palmer asked Kamisha.

The instructor pushes the prospective teacher's thinking by asking questions, not by suggesting answers.

"It seems like she is only using the numerators to determine which of them is more. "

"What difference does that make?" asked Dr. Palmer.

"She's not thinking about how much of the denominator it is."

The instructor asks questions that encourage the prospective teachers to call upon their own mathematical knowledge to assess the depth of student understanding.

"What do you mean? Lisa asked. "She seems to understand that the denominator is the number of things in the set."

"But when she's comparing two fractions of different-sized sets she is only considering the number of parts and not how much of the total set they are."

The university instructor uses her students' observations of children's partial understandings to highlight difficult concepts and enrich her students' understandings of the topic.

"That is one of the things that makes fractions complex," Dr. Palmer commented. "Not only are we interested in the numerator and denominator, but we also want to know about their relationship to one another."

"Perhaps she is thinking about that, but she is considering two different-sized sets," Lisa suggested.

"Could you say more about that, Lisa?" Dr. Palmer asked.

The thinking of each prospective teacher is encouraged.

"One of the things Latalya is comparing is twice as big as the other and all the pieces are the same size. So two of the larger set *is* more than one of the smaller set." Lisa went to the board and drew boxes around the circles in Latalya's drawing to show that the set of hamsters was twice as large as the set of dogs.

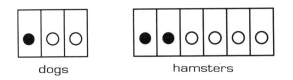

"Her ²/₆ is more than her ⅓," she concluded.

"Tia is doing something different," Peter volunteered. "She is comparing two equal-sized sets. She has just divided them up unequally. She may not realize that each thing should be divided into equal pieces, although she seems to understand the part-whole relationship."

Turning to Mr. Konook, Damon suggested, "Maybe the task should have already had the pictures drawn with it; then these problems wouldn't have come up."

"Actually," explained Mr. Konook, "I intentionally put the problem on the board as a symbolic statement and asked them to show how they got their answers. I wanted to see what they understood about comparing fractions. Listening to your descriptions of the conversations you had with these students has been very helpful."

"Wow," exclaimed Damon, "I was thinking that you would want to give the students a model to use to help them get the answer and avoid the confusion. But it seems like having them come up with their own models and allowing for some confusion revealed what the students do and don't understand. These two students got the same answer, but for very different reasons."

"It seems like thirds and sixths are hard to work with," Damon noticed. "They are much more obscure than halves and fourths."

"That's a good point. Why did you decide to use thirds and sixths in this problem?" Lisa asked Mr. Konook.

"You're right, Damon, about thirds and sixths being more foreign to young children. Up until this point we've mostly worked with quarters and halves. The students have a lot of informal knowledge about halves and a little bit about fourths. We used their knowledge of how to write ½ to determine the meaning of the numerals within a fraction. They were able to connect this knowledge to their knowledge of quarters pretty readily because they all knew that there are four quarters in a dollar. They are not as familiar with thirds and sixths, so I was interested in how they would use their understanding of halves and fourths in working with thirds and sixths."

Dr. Palmer encouraged her students to consider with Mr. Konook what the students understand about thirds and sixths. She followed this by asking them how they would further develop these ideas in the next day's lesson.

Ways to think about fractions come out naturally in the discussion. The prospective teacher is disposed toward understanding the students' thinking rather than merely judging it as right or wrong.

The prospective teachers identify a critical concept and then analyze the instructional approach taken by the teacher.

The classroom teacher's analysis of the task and rationale for using it model the process of selecting and posing worthwhile tasks. He also uses the open-ended situation to assess children's thinking.

This exchange shows professional collegiality in analyzing and evaluating effective teaching practices. Through this interaction, the prospective teacher examines and revises his assumptions.

Students are encouraged to consider subtle aspects of teaching fractions.

A knowledge of children's experiences influences what the teacher does. The prospective teachers are encouraged to pay attention to students' thinking and to make pedagogical decisions based on their knowledge of the students.

The instructor pushes the students to think about the next step in developing children's understanding of fractions. The entire activity focuses on analysis as an ongoing part of teaching. Dispositions to be a reflective teacher are enhanced.

The instructor considers what she knows about prospective teachers in making decisions about the curriculum for this introductory course.

The instructor focuses on ways to help the teachers examine and revise their assumptions about mathematics, how it is taught, and how students learn it..

The instructor models good teaching of mathematics as she engages the prospective teachers in learning mathematics.

Her students are challenged by the content and the new way they are expected to engage in learning.

The instructor demonstrates that children can learn about permutations, which, on the surface, may appear to be a relatively complex topic to the prospective teachers.

This is an opportunity for the prospective teachers to examine and revise their notions about how children think and what they know. Teaching someone else addresses the prospective teachers' assumptions about how mathematics should be taught.

Even though the prospective teachers have experienced and observed the use of new teaching strategies, it will take time before they can assimilate

5.2　Janet Foote, a mathematics educator at a small college, has been trying to find ways to engage her students in the many aspects of what it means to teach mathematics well. Each term, she confronts the same issues. Her students are convinced that mathematics will be hard for them—they believe that they are not "mathematically minded." They also believe that learning mathematics is synonymous with being *told* information and that mathematics is learned through repetition. Indeed, the teacher's job is to *tell* students the mathematical knowledge they need to know. Dr. Foote's students expect the same "opportunity"—they want her to *tell* them how they should explain mathematics to children.

Dr. Foote wants to help these students learn some mathematical content and, at the same time, reconsider their assumptions about what it means to learn and teach mathematics, particularly as it deals with the crucial pedagogical skills of analysis, questioning, and creating tasks.

This fall, Dr. Foote decides to try a different approach. She designs a three-phase learning cycle. First, to give her students a new experience with being a learner of mathematics, she has them grapple with the topic of permutations, since this is a topic that many students are unlikely to have studied, or if they have, they are unlikely to understand its conceptual basis.

Exploring the problem of how many different license plates can be made in their state (which has license plates with three letters and three numbers), Dr. Foote concentrates on helping her students learn mathematics in a way that is quite different from their earlier experiences. She does not show them what to do. Rather, she asks probing questions about their thinking and about the conclusions they have drawn. She deliberately seeks neither to approve nor to disapprove of their answers, expecting them to prove their answers through argument and justification.

During the two class periods, she involves them in a variety of related problems. As one of her students notes: "When I learned permutations in high school, I was just expected to use the formula for finding the factorial of the number...just carry out the multiplication. I didn't understand what was behind the formula. When I have to think about *why*, then I get confused. I have to develop a new way of investigating."

For the second phase of the learning cycle, Dr. Foote brings three second-grade students to class. With the prospective teachers observing, she does a series of related problems with the children. In one activity, the children explore ways to form one, two, and three digit numbers by lining themselves up in different orders.

The prospective teachers' focus during this time is on the ways Dr. Foote interacts with the children, the tasks she selects, and the ways the children respond. At the end of this session, she encourages the prospective teachers to ask the children any questions they may have. Following this, in the next class session, she and the prospective teachers discuss the teaching and learning that occurred.

As the final phase, the students become teachers—they are expected to try to help someone else explore the concept of permutations. "Someone else" may be children, roommates, parents, and so on. They make preliminary preparations during class, meeting together in small groups.

The discussion that follows this phase is enlightening. As Dr. Foote suspects, the students begin to realize that there is more here than meets the eye. In particular, the students struggle with wanting to *tell* their learners instead of helping them build their own understandings. As one student notes:

"I tried to teach my roommate something about permutations. It just didn't work. I thought I understood permutations enough myself, but when she began asking me questions, I was lost. I couldn't answer some of them, which made me think that I really didn't understand as well as I thought I did."

5.3 Ms. Costa, the director of a local state-supported teachers' project, wants to encourage participating teachers to examine their assumptions about the nature of mathematics and how it should be taught. In particular, she wants them to appreciate the difference between understanding how to solve a problem and merely being able to apply an algorithm. At a meeting she poses this problem:

> In a certain adult condominium community, two-thirds of the men residents are married to three-fifths of the women residents. What part of the community is married?

Bob: There isn't enough information.

Tony: Are the people in the community only married to others in the community?

Connie: We don't know how many people there are. Are there the same number of men and women?

Ms. Costa: You raise some good issues and questions. What are some conditions that aren't explicitly stated but that we would assume in order to make sense of the problem?

Tom: Men and women in the community who are married are only married to others in the community.

Connie: There are the same number of men as women.

Susan: I don't think that's true. The number of married men should be the same as the number of married women but the totals could be different.

The teachers continue to list several assumptions, selecting those with which they agree and recognizing others as misleading.

The teachers work on the problem in small groups while Ms. Costa walks around the room. She notices that some people are still clarifying issues raised in the initial discussion while others have agreed on conditions and are trying to determine what part of the community is married.

A variety of approaches emerge from the small-group discussions. The participants later share these with the class. They are encouraged to generalize: How are the solution strategies alike and how are they different? Do they all produce the same result? What if the fractions in the original problem are changed?

Ms. Costa: Now suppose this sort of problem is included as part of the school mathematics curriculum. I'd like to demonstrate a possible way to present the problem and solution to students.

Ms. Costa then walks the teachers through a rule-based approach to solving the problem. She develops a three-step procedure and explains exactly what calculations should be done at each step.

such changes into their own personal views of what constitutes good teaching.

The student recognizes that her own knowledge is incomplete. At the same time, she assumes that the role of the teacher is to answer questions— it is hard for her to consider that questions may be used to guide the investigation.

The project director has selected a problem that is not typical of those that appear in most textbooks. She expects it to raise many questions among the teachers.

In order to solve the problem the teachers make explicit their assumptions and determine which are necessary to its solution.

The instructor allows participants time to grapple with the problem and discuss their ideas with each other.

The members of each group select and use a strategy that makes sense to them and eventually analyze a range of strategies for approaching the problem.

The instructor invites the teachers to reflect on what they know about the traditional mathematics curriculum and the ways of presenting new kinds of problems.

Step 1. Find the L.C.N. (least common numerator). [The L.C.N. for 2 and 3 is 6.]

Step 2. Change fractions to equivalent fractions with same L.C.N. [2/3 = 6/9 and 3/5 = 6/10.]

The final step is to use the "Add Add" method of combining fractions. In this procedure we combine two fractions by adding the numerators to get the new numerator and the denominators to get the new denominator.

Step 3. Add Add the two fractions. [6/9 + 6/10 = 12/19.]

Ms. Costa: These three simple steps will allow you to solve all "marriage type" problems.

Ms. Costa visually surveys the participants to observe their reactions. Some participants seem surprised, others confused, and still others are nodding in agreement with the answer and procedure.

The instructor expects reactions. Many of the teachers may be experiencing some dissonance between how they solved the problem and how students are likely to be taught to solve such problems.

Several "good" arguments are provided to justify the "tell, show, and do" model of problem solving. The instructor asks the teachers to discuss the differences they observed.

The instructor helps a teacher recall that although the steps may make sense now, they probably were not the focus during the group's initial efforts.

The goal is for the teachers to realize that "easier" is not necessarily better when it comes to truly understanding mathematics.

Ms. Costa: What do you think? It didn't take very long and could be accomplished quite easily in the school curriculum. Of course you would have to spend time memorizing the three steps and practicing them. How does the method compare to the methods you used earlier to solve the problem?

Joe: It's quick, but it doesn't explain what you're doing.

Marissa: It was really the same as my group did before, only more formal. We found the same numerator. Then the total number of each is the denominators. But we just counted everything to get the totals.

Ms. Costa: Your group tried to make sense of the three steps. Do you think you could have done that without having had time to work on the problem on your own and comparing the various methods we came up with before?

Tony: I don't think so but it sure seems easier for the kids to follow the three steps.

Ms. Costa: Well, I could have started today's lesson by presenting the problem and giving you the three steps to solve it. We could have spent time practicing the steps and doing similar problems. You could have been quite successful in solving problems of that type without understanding much about why you were doing the things you did. In the short run you would appear to be successful, but in the long run where would you be?

Cross-Referencing Standard 5 with Vignettes in Section II

The vignettes presented in the second section, "Standards for the Evaluation of the Teaching of Mathematics," reflect teaching episodes documenting evaluation as professional development and assessing the teaching of mathematics. As might be expected, these vignettes demonstrate teachers' applications of the components detailed in this standard as well. The chart below provides a cross reference from the subtopics of this standard to the vignettes presented in the second section. For example, vignette 1.2 provides an example of a teacher examining and revising his assumptions about how mathematics should be taught.

As part of the preservice and continuing education of teachers of mathematics, the vignettes from the second section may be helpful as part of teachers' efforts to think about their own teaching in light of the subtopics of this standard.

Standards for the Evaluation of the Teaching of Mathematics - Vignettes

	Vignette number and page number															
	1.1 76	1.2 77	1.3 78	2.1 80	2.2 82	3.1 85	3.2 86	4.1 90	4.2 91	5.1 96	5.2 98	5.3 100	6.1 104	6.2 106	7.1 111	7.2 112
◆ examine and revise assumptions about the nature of mathematics, how it should be taught, and how students learn mathematics;		●			●									●		
◆ observe and analyze a range of approaches to mathematics teaching and learning, focusing on the tasks, discourse, environment, and assessment:		●	●	●				●	●		●	●			●	●
◆ work with a diverse range of students individually, in small groups, and in large class settings with guidance from and in collaboration with mathematics education professionals;	●										●	●		●	●	●
◆ analyze and evaluate the the appropriateness and effectiveness of their teaching;	●		●	●	●			●	●	●				●	●	●
◆ develop dispositions towards teaching mathematics.					●								●	●		

STANDARD 6:
THE TEACHER'S ROLE IN PROFESSIONAL DEVELOPMENT

Teachers of mathematics should take an active role in their own professional development by accepting responsibility for—

♦ *experimenting thoughtfully with alternative approaches and strategies in the classroom;*

♦ *reflecting on learning and teaching individually and with colleagues;*

♦ *participating in workshops, courses, and other educational opportunities specific to mathematics;*

♦ *participating actively in the professional community of mathematics educators;*

♦ *reading and discussing ideas presented in professional publications;*

♦ *discussing with colleagues issues in mathematics and mathematics teaching and learning;*

♦ *participating in proposing, designing, and evaluating programs for professional development specific to mathematics;*

♦ *participating in school, community, and political efforts to effect positive change in mathematics education.*

Schools and school districts must support and encourage teachers in accepting these responsibilities.

Elaboration

Teachers develop as professionals on an ongoing basis. Focusing on their classroom practice, they experiment with alternative approaches to engage students in mathematical ideas, possible strategies for assessment, and different ways of organization. They analyze and adapt strategies that they try, examining how well they help students develop mathematical competence and confidence. They incorporate such strategies into an ever-growing and more complex repertoire. Beyond the classroom walls, teachers also evolve as participants in a wider educational community. They read, talk with colleagues, take the initiative to press for changes, and raise their voices to speak out on current issues. Teachers' professional development, within and outside their classrooms, is a product of their reflectiveness and participation in educational opportunities that will enhance and extend their growth and development. As professionals, teachers take responsibility for their own growth and development.

There is a voice that is not heard often enough in schools these days: the concerned voice of the informed mathematics educator. We invite you to develop this voice. Having it, you can and should become an authority figure in your school—maybe not a power figure, in the sense that a principal has power—but an authority nonetheless. Your authority will come from knowing the things about the teaching and learning of mathematics that can be clearly known—knowing what is being tried about the country and with what success, knowing current opinions on what ought to be done, knowing your own program from stem to stern, and, above all, knowing the questions that one must keep asking. (Ohio Mathematics Education Leadership Council 1989, p. 1)

In addition, professionalism among teachers is built through a support system that links them to colleagues inside and outside the schools. Teachers should be able to turn to colleagues for information concerning any aspect of mathematics education in order to expand their views of mathematics, their resources for teaching, and their repertoire of teaching and learning skills. Such interchange provides intellectual refreshment and places teachers in the role of partners in the process of education. It also provides opportunities for heightened awareness of the responsibility for fostering their own professionalism by building collegial networks, reading professional literature, becoming involved with professional organizations, and initiating contact with teacher educators at local colleges and universities.

Teachers can take an active role in their professional development through such activities as—

♦ forming special-interest groups within their schools to investigate ways technology might better enhance their teaching;

♦ participating in summer programs to learn new topics in mathematics such as statistics or discrete mathematics;

♦ meeting with teachers from neighboring school districts to explore how they can jointly offer advanced mathematics courses for their students via telecommunications;

♦ working on curriculum renewal with other mathematics faculty to change the nature and kinds of courses that are being offered and align their program with the *Curriculum and Evaluation Standards*;

♦ joining local mathematics associations, attending meetings, making presentations, and assuming leadership roles.

Teachers who are engaged proactively in making mathematics education better demonstrate this in many ways. What is essential is that they view themselves as agents of change, responsible for improving mathematics education at many different levels: the classroom, the school, the district, the region, and the nation.

Vignettes

6.1 Dick Richey is a mathematics teacher in a large high school. Five years ago he was disenchanted with teaching. He was bored, his students were bored, and there did not seem to be any challenge in the job. He even considered leaving teaching.

He read about a three-year, federally funded summer institute on the teaching of algebra in an issue of the NCTM *News Bulletin* in the faculty resource room. He decided to apply but was rather apprehensive when he was selected.

Professional publications can help teachers keep in touch with the mathematics education community. In addition to providing substantive ideas about mathematics teaching, they can be sources of information about professional development opportunities.

During the first three-week summer session he was immediately thrust into the midst of twenty-four experienced teachers. Two university professors were determined to involve him in thinking deeply about what he was doing in his classes. Some of the mathematics content in the sessions was new to him. Other sessions focused on using technology to teach mathematics. Still others centered on teaching strategies and research on teaching and learning. Sharing and collegiality dominated their work together.

Recognizing that significant teacher change requires ongoing support over time, the institute planners have developed a three-year program.

Dick remembers he was initially convinced that the professors were

Analysis of the curriculum with colleagues encourages the teachers to examine their own teaching and leads them to change the way they think about teaching.

The teachers' knowledge of mathematics is enriched through conceptualizing its organizational structure.

Reflecting on teaching and learning leads to changes in teaching practice.

Through designing and participating in professional development opportunities the teacher and his colleagues contribute mutually to one another's growth.

Continued contact and collaboration over an extended period of time support the teachers in changing their practice.

Contributions to the professional community become a regular part of the teacher's career efforts.

Working together, the teacher, the principal, and a local business effect positive change in mathematics instruction at both the school and district levels.

The network within the mathematics education community has been a resource as the teacher plans professional development activities for other teachers.

Stacy is disposed to reflect on and analyze her teaching from the perspective of what students should learn and are learning. Teachers can begin to develop such a disposition in preservice teaching programs.

unrealistic. They gave an assignment to the group to analyze the first-year algebra curriculum. The teachers were to identify the "big" ideas of algebra—that is, the ideas that were so powerful that understanding them would enable every student to do any beginning algebra problem. The trick was that they were permitted to choose no more than ten "big" ideas!

When the teachers pooled their lists of "big" ideas, it was three pages long and looked just like the table of contents of a first-year algebra text! Over the three-year period the teachers developed lenses to look at the algebra they were teaching to help them identify central ideas. Two summers later the list had been refined to a very small but significant list: real numbers, variables and functions, distributive property, equivalent fractions, and expressions and sentences.

Thinking about algebra as being organized around several key ideas made a considerable difference in Dick's teaching. He found himself looking for connections among ideas and trying to help his students find mental hooks or organizers on which to hang new mathematical ideas. In addition, he occasionally videotaped his classes so that he could analyze his teaching.

Dick was terrified when he had to "go public" for the first time. He wasn't sure that what he was trying to do in his classes was anything new. Much to his surprise other teachers seemed to be very interested and gave him some other new ideas.

The continuing support and yearly visits from the professors helped Dick maintain his renewed perspective on his teaching. Throughout the program, the institute participants shared their ideas and struggles through electronic mail. In fact, this network continued and grew to include a great many high school teachers across the state.

In the years since the institute, Dick has continued to grow as a professional. Subsequently, he has been chosen to participate in additional summer programs. He and a colleague recently published a manual on teaching with graphing calculators.

The support Dick has received from his principal and the district mathematics supervisor has been essential to the changes Dick has made. They have provided him with time to work with the other faculty in the district and to write a "Technology in Education" grant proposal to a local business. The grant has provided the district with resources to buy sets of calculators and to upgrade the computers that are available for mathematics instruction.

Although the algebra institute is over, Dick has remained in contact with the professors and many participants, personally and electronically. Their support has been particularly helpful in planning workshops that he has given in the area.

6.2 Stacy Washington, a fourth- and fifth-grade teacher for three years, has grown increasingly dissatisfied with her mathematics teaching. She feels that she has successfully made the other subjects she teaches come alive for her students, encouraging them to think for themselves and engaging them in group discussions.

Stacy believes that her students should understand the mathematics they are doing, learn to reason mathematically, and use this knowledge to solve problems. However, they must spend so much time working alone to learn the mathematical rules and procedures that there seems to be no time for lively group discussions. She wonders if there is some other way to help her students learn the material that would still leave time for stimulating group work.

Stacy has tried to use concrete materials as tools in her teaching, but she feels frustrated by the manipulatives she has used. Topics such as long division and decimals don't lend themselves readily to the use of manipulatives. She did use fraction bars when teaching addition and subtraction of fractions, but they were just tools to help get the right answer. In addition, that particular representation completely broke down for her when she introduced multiplication of fractions.

Determined to get help, Stacy expresses her concerns to her principal, LaTasha Enary-Fayse, who suggests that she talk with teachers who share her concerns.

After making a few phone calls LaTasha gives Stacy the name of a teacher in a neighboring district who has made changes in his math teaching over the last few years. She agrees to arrange for a substitute once Stacy has scheduled a visit.

Stacy's visit with Jon Nickerson proves to be fruitful. Not only does she observe his mixed-ability seventh-grade students engage in sophisticated and serious dialogue about mathematics, but it is clear that they are developing basic skills through discovering patterns, articulating them to the class, and determining whether or not they could be generalized.

While comparing the graphs of four different linear equations on the computer, one student noticed that the two lines with "the same number in front of the x were parallel."

The students have not been taught the definition of slope, but they are discovering patterns that relate to the slope of the lines. Jon asks questions like, "Will this always be true? How do you know? Do you agree? Why?"

Stacy tries to imagine her students in a similar discussion. She realizes that she knows little about what her students are thinking about the mathematics they are learning.

Afterwards Stacy has the opportunity to talk to Jon. He understands her concerns about her mathematics teaching and reflects on his own experience.

As a mathematics major, he knows firsthand that learning mathematics involves more than memorizing rules and practicing them, but it took some experimenting to foster the sort of inquiry and discourse that Stacy had observed in his classroom. "In fact, I am always learning more and trying out new ideas," he explains. "Getting my master's degree in mathematics really helped me think about the relationship between advanced mathematics and what I teach in middle school."

Jon shows Stacy his copy of the NCTM *Curriculum and Evaluation Standards for School Mathematics* and encourages her to join NCTM to take advantage of their regular publications. As they look through it together, she is struck by the emphasis on understanding that goes beyond getting right answers. They discuss several strategies that Stacy might use in her class to start things off.

The teacher uses her knowledge of the nature of mathematics to contemplate appropriate school mathematics.

Alternative approaches are not used indiscriminately, but are thoughtfully considered, tried, and carefully analyzed.

The teacher pursues her own development through seeking help from an administrator.

The administrator facilitates collegial support by assisting in making contact and providing release time for the teachers to meet.

Cross–grade-level observations and collaborations can contribute mutually to the teachers involved.

The teacher observes basic skills being developed and practiced in problem-solving tasks.

The teacher asks questions that challenge his students' thinking and encourage them to listen to one another and, as a group, makes sense of the mathematics.

Through observing a colleague, the teacher reflects on her own teaching.

Jon's teaching is not composed of techniques that Stacy can assimilate through a single observation. Following up with a discussion of issues and concerns provides Stacy with insights into the decision making underlying his teaching.

Professional development continues to be an ongoing process for this experienced teacher.

Through participation in professional organizations teachers can contribute to their own development.

Stacy leaves convinced that her students should explain their answers and that they should learn to listen to, and make sense of, other students' solutions. She and Jon agree to meet regularly to share their progress and struggles. They plan to invite their colleagues to join them in these discussions. Jon also offers to observe Stacy's class when she feels ready.

Throughout the rest of the year, Stacy's progress is gradual and steady. Eventually Stacy is comfortable having Jon and LaTasha visit her classroom. LaTasha also asks if she could arrange for other teachers to visit some of Stacy's math lessons.

Toward the end of the year LaTasha asks Stacy if she would be willing to work with other mathematics teachers in the district to make curriculum decisions and plan professional development opportunities. She happily agrees and is looking forward to learning from the other teachers. She is surprised to realize that she had never thought of herself as a mathematics teacher before, even though she teaches it everyday.

6.3 Dr. Jackson is coordinator for a field-based mathematics education course for elementary school teachers at a state university. For a number of years, she has experienced a great deal of difficulty in scheduling local elementary classrooms for the necessary field experiences. There are usually five or six sections offered each semester that meet at different times. Locating a sufficient number of classrooms and avoiding overcrowding present a challenge.

Before the start of the semester, Dr. Jackson raised her concerns with Dr. Pruitt, the principal of a local elementary school. Dr. Pruitt agreed that the field experience should be an integral part of the prospective teachers' program and promised to talk with his teachers.

At school, Dr. Pruitt shared Dr. Jackson's concerns with his staff. The teachers agreed that field experiences are essential to professional education. They selected a representative from each grade level to meet with Dr. Jackson to discuss the matter.

At the meeting, the teachers wanted to know what kinds of field experiences Dr. Jackson had in mind. She indicated that she had two major goals: to have her students observe the kind of mathematics class being recommended by the *Curriculum and Evaluation Standards* and to have students conduct one-on-one clinical interviews in order to have the experience of analyzing a child's thinking. The teachers were excited about the clinical interviews, since it would provide them with another assessment resource.

Together, the teachers and Dr. Jackson developed the following plan. For five weeks during the semester, probably during the seventh through the eleventh weeks, the teachers will rearrange their schedules once a week so that math time will fit Dr. Jackson's course times. They would also reserve another week for the interviews.

Before the meeting ended, one teacher indicated she was interested in having university students teach small groups of students some mathematical activities. Dr. Jackson was enthusiastic, since it was compatible with one of the course assignments—creating mathematical task files. The request was well received by the other teachers. They decided that the last three classroom visits would be used for this purpose.

Teachers can grow professionally through ongoing participation in the community of mathematics educators on local as well as broader levels.

The collegial interaction involved in classroom observations and subsequent discussions can contribute to the professional development of both teachers.

Participation in school and district efforts to effect change in mathematics education provides teachers with opportunities to develop professionally through interaction with colleagues,

The coordinator recognizes the problems inherent in developing opportunities for prospective teachers to participate in field placements.

The principal supports field placements but recognizes that his teachers need to be involved in any planning that will include students in their classrooms.

The teachers recognize that they need to participate in their professional community, in this case, helping future colleagues participate in fruitful field experiences.

Such involvement helps the teachers in their efforts to experiment with alternative strategies for assessing their students.

By their involvement, the teachers are participating in schools' efforts to effect positive change in the mathematics education of preservice teachers.

Teachers help Dr. Jackson refine her program, enhancing opportunities for her students to gain experience working with small groups of children.

After the meeting, Dr. Jackson and Dr. Pruitt discussed how they might help each other. Dr. Jackson expressed an interest in having some teachers participate as guest speakers because the prospective teachers valued the opinions of practicing teachers. Dr. Pruitt agreed to discuss this proposal with the district office, hoping to give teachers release time to visit Dr. Jackson's classes.

In addition, Dr. Jackson also indicated that she would like to explore ways that she could do some teaching, perhaps teaming with one of the teachers for a few months in planning and teaching mathematics. Dr. Pruitt was delighted and suggested that they plan another meeting to discuss this further.

The teachers are being asked to participate in educational opportunities that permit them to share their expertise with preservice teachers.

STANDARDS FOR THE SUPPORT AND DEVELOPMENT OF MATHEMATICS TEACHERS AND TEACHING

OVERVIEW

This section presents four standards for the support and development of teachers and teaching:

1. Responsibilities of Policymakers in Government, Business, and Industry

2. Responsibilities of Schools

3. Responsibilities of Colleges and Universities

4. Responsibilities of Professional Organizations

INTRODUCTION

The *Professional Standards for Teaching Mathematics* presents a vision of teaching that calls for a teacher who is educated, supported, and evaluated in ways quite different from current practice. To create teaching environments that encourage mathematical problem solving, communicating, reasoning, and connecting ideas—in short, mathematical inquiry and decision making—teachers must have access to educational opportunities over their entire professional lives that focus on developing a deep knowledge of subject matter, pedagogy, and students. Many external forces and decisions affect mathematics teaching and school mathematics programs. Various constituencies have responsibility for the support of mathematics teachers and teaching and in building successful mathematics programs. Such support needs to be multifaceted, systemic, and reliable.

Existing support systems for mathematics teachers are as inadequate for teaching in today's society as the shopkeeper arithmetic curriculum is for educating our children to live and work in the twenty-first century. The kind and level of mathematics education required for today's students to prosper in a dramatically changed economy and in a scientifically and technologically advanced society places great responsibility on the shoulders of teachers of mathematics. At the same time our society is undergoing other dramatic changes that make teaching even more challenging. We are growing more diverse along many dimensions—ethnically, culturally, linguistically, in family patterns, in the integration of persons with disabilities into mainstream institutions, and in numerous other ways (Secada 1990).

Teachers can and do implement successful mathematics programs with little help or encouragement. However, such practice should not be expected to flourish without adequate support. The changes called for by the *Curriculum and Evaluation Standards* and the *Professional Teaching Standards* need the support of policymakers in government, business, and industry; school administrators, school board members and parents; college and university faculty and administrators; and leaders of professional organizations. Each of these individuals and groups has responsibilities to help support and shape the environment in which teachers teach and students learn mathematics. The standards in this section focus on these responsibilities.

STANDARD 1:
RESPONSIBILITIES OF POLICYMAKERS IN
GOVERNMENT, BUSINESS, AND INDUSTRY

Policymakers in government, business, and industry should take an active role in supporting mathematics education by accepting responsibility for—

◆ *participating in partnerships at the national, state or provincial, and local levels to improve the teaching and learning of mathematics;*

◆ *supporting decisions made by the mathematics education professional community that set directions for mathematics curriculum, instruction, evaluation, and school practice;*

◆ *providing resources and funding for, and assistance in, developing and implementing high-quality school mathematics programs that reach all students, as envisioned in the* Curriculum and Evaluation Standards for School Mathematics *and the* Professional Standards for Teaching Mathematics.

Elaboration

Policy decisions are made at many levels that affect the status of teachers and teaching and the environment in which teachers teach mathematics. High school graduation requirements, state-mandated pupil and teacher testing, state department of education certification requirements, textbooks published and adopted, standardized tests published and adopted, local scheduling of classes, teaching assignments, allocation of resources, policies that affect professional development such as attendance at professional meetings, teacher evaluation procedures—these are but a few of the myriad decisions that either lend support to or substantially constrain the improvement of mathematics teaching and learning. All too often such policy decisions are made without the consultation of teachers and other mathematics education professionals; yet it is the mathematics education community whose education, experience, and expertise must be reflected in making wise policy decisions that affect developing mathematical power for every student. Teachers at all levels are held accountable for the mathematical growth of students. Therefore, to be effective, policymakers in all arenas—state, provincial, local, national, business, and industry—must confer with and support teachers and other mathematics education professionals on issues that affect what and how a teacher of mathematics can teach.

Businesses that have a particular responsibility are publishers of textbooks and standardized tests. We know that textbooks and tests have a profound influence on what is taught. Therefore, authors and publishers have both an opportunity and a responsibility to help improve mathematics instruction. Publishers should seek advice from teachers and other mathematics educators in making decisions that affect what is developed and published in mathematics. Teachers have a responsibility to insist on materials that are the best suited for children in learning mathematics and on testing that is aligned with the goals of the mathematics program.

Helping students develop mathematical power takes tools, resources, and experiences very different from previous practices. Policymakers in government, business, and industry must understand the need for materials and tools for learning and doing mathematics and must join together in partnerships and coalitions to find ways to provide appropriate resources for teaching mathematics. There are many examples of such partnerships. Computer companies have designed programs that have made computers available for school use in some areas. Groups have formed within communities and at the state or province level to help provide resources for schools.

Business and industry policymakers should recognize that they stand to gain very substantially from excellent school mathematics programs. Huge amounts of money are spent each year to train workers in the mathematics needed in the technological workplace. Having a work force that is better educated in mathematics in their K–16 programs would change the nature of on-the-job training and have an impact on economic competitiveness. Therefore, business and industry must join with communities and schools in improving mathematics instruction. The kinds of help needed are both financial and human. Allowing scientists, mathematicians, and other employees to spend time in schools can be very beneficial in stimulating students to study more mathematics, especially when these business and industry personnel are culturally diverse. Interacting with African American, Hispanic, American Indian, female, or disabled scientists helps both students and teachers set higher expectations for all students. Setting up programs that allow teachers to have internships in business and industry can stimulate teachers' professional growth and help them bring the reality of the uses of mathematics in the workplace to their students. This also sends signals that mathematics teaching and teachers are valued in our society, thus making mathematics teaching a more attractive career option for talented young people.

Policymakers must understand the mathematical needs of workers and citizens of the future and join together to help make such a mathematics education a reality for all students. If teachers are to be able to realize such a goal in the classroom, they must have the financial and other support to continue to learn. They must have the time to reflect on their teaching and on students' understandings so that all students can be reached. This means dramatic restructuring of teaching assignments and provision for ongoing professional development opportunities for teachers of mathematics. Such restructuring can only be accomplished with the support and involvement of policymakers in government, business, and industry. Schools, teachers, and students—in fact, our society—need and deserve such help.

STANDARD 2:
RESPONSIBILITIES OF SCHOOLS
AND SCHOOL SYSTEMS

School administrators and school board members should take an active role in supporting teachers of mathematics by accepting responsibility for—

◆ *understanding the goals for the mathematics education of all students set forth in the* Curriculum and Evaluation Standards for School Mathematics *and the needs of teachers of mathematics in realizing these goals in their classrooms;*

◆ *recruiting qualified teachers of mathematics, with particular focus on the need for the teaching staff to be diverse;*

◆ *providing a support system for beginning and experienced teachers of mathematics to ensure that they grow professionally and are encouraged to remain in teaching;*

◆ *making teaching assignments based on the qualifications of teachers;*

◆ *involving teachers centrally in designing and evaluating programs for professional development specific to mathematics;*

◆ *supporting teachers in self-evaluation and in analyzing, evaluating, and improving their teaching with colleagues and supervisors;*

◆ *providing adequate resources, equipment, time, and funding to support the teaching and learning of mathematics as envisioned in this document;*

◆ *establishing outreach activities with parents, guardians, leaders in business and industry, and others in the community to build support for quality mathematics programs;*

◆ *promoting excellence in teaching mathematics by establishing an adequate reward system, including salary, promotion, and conditions of work.*

Elaboration

The *Curriculum and Evaluation Standards for School Mathematics* and the *Professional Standards for Teaching Mathematics* lay out a vision for school mathematics and instruction that should become the framework for planning school mathematics programs. It is essential that principals, superintendents, and other administrators understand these documents so that they can represent the mathematics program to the community, especially to parents, in ways that help generate the support teachers need to implement a high-quality mathematics program that meets the needs of every student. School administrators must understand that the mathematics education called for in these standards requires time for mathematics; time for teachers to plan, to reflect, to help each other improve instruction; time for professional development; and time to interact with the community. The payoff for such commitment is the realization of a mathematics program that fosters the development of mathematical power for all students.

As our society grows more diverse ethnically, culturally, linguistically, and in the integration of persons with disabilities into mainstream institutions, the teaching force must also change to reflect this diversity. For example, having a school environment that includes teachers of mathematics who are American Indian, African American, or Hispanic will help children of such ethnic groups see themselves as capable members of society who can do significant mathematics. To achieve this diversity among teachers of mathematics requires aggressive, systematic, and long-term support for the young people who are recruited into the mathematics teaching profession.

This should begin with the school community itself. If the administration, the school board, and the parents show their respect and support for teachers and teaching as a profession, then talented young people are more likely to consider teaching as a career choice. Once in the profession, beginning teachers need supportive guidance to help them develop their skills and habits of mind as teachers of mathematics. Such beginning teacher programs are essential to retain talented young people in what is a very demanding, as well as rewarding, profession.

The teaching of mathematics, like any other subject, requires knowledge and experience that is specific to that discipline. Although one may talk about tasks, discourse, environment, and analysis in relation to any school subject, the knowledge of how students learn, of the subject matter content, and of the pedagogy are specific to that discipline. An excellent social studies teacher should not be assigned a section of mathematics under the assumption that his or her teaching knowledge and skills will transfer. Teaching assignments must be made on the basis of the qualifications of teachers.

Schools have major responsibilities for fostering the professional development of teachers of mathematics. Their work in supporting appropriate professional development programs, promoting collegiality, and recognizing the role of teachers as responsible professionals does much to enhance quality programs and instruction in mathematics education. Schools should allocate a fixed percent of their budgets for faculty development activities.

The unique challenges facing teachers of mathematics are best addressed when they are provided with opportunities to engage in ongoing, subject-specific, professional development programs. Their concerns and interests are met when teachers play a major role in identifying and assessing their own needs. Planning and developing continuing education programs should take place within individual schools and school districts and should highlight teacher involvement both in program development and program implementation.

Professional programs require the commitment of resources, equipment, time, and funding. For example, the ready availability of technology in the workplace and the implications of that resource for use in mathematics teaching and learning underscore the critical need for including this equipment as instructional and learning tools.

As a part of professional development, principals should allocate time for teachers to build collegial links with other faculty. Teachers need these opportunities to share ideas, plan interdisciplinary lessons, and explore instructional strategies. Returning to the university to pursue graduate programs in mathematics education allows teachers to deepen their knowledge and skills in mathematics and in mathematical pedagogy. Schools that are supportive of such activities for teachers will benefit

from a teaching staff that continues to learn and to improve the mathematics program and students' successes in mathematics.

Mathematical power for our students cannot be developed with chalkboards and ditto sheets, inadequate class time, and six classes a day per teacher. Mathematics teachers need appropriate resources. Calculators, computers, and manipulatives are as integral to learning and doing mathematics as chemicals are to a chemistry laboratory. If mathematics teachers are to manage the growing diversity in their classrooms, time to plan, to study, to reflect, to develop curriculum, to confer with colleagues is essential.

Mathematical power must be a concept to which we commit for *all* students, not just for the privileged few. This requires a commitment from the school and community for adequate funding to support the teaching and learning of mathematics. However, in order for the community to be supportive of the mathematics program they must know what the program goals are and must understand the kind of support needed by teachers to carry out the program. Here the school administration, especially the principal, is key. Principals who take the time to work with their teachers in developing a coherent, powerful mathematics program and, further, take the responsibility to be the advocate for the teachers to the community can make it possible for teachers to teach and for students to learn as envisioned in the *Curriculum and Evaluation Standards for School Mathematics* and in these *Professional Standards for Teaching Mathematics*.

In order for teachers to move toward the vision of teaching in these standards, school administrators must establish a reward system, including salary and promotion, that supports and encourages teachers as they grow professionally. One productive and rewarding way to support teachers in making improvements in the mathematics program is to fund extended year contracts for summer pay to develop special projects that teachers have proposed.

STANDARD 3:
RESPONSIBILITIES OF
COLLEGES AND UNIVERSITIES

College and university administrators need to take an active role in supporting mathematics and mathematics education faculty by accepting responsibility for—

◆ *establishing an adequate reward system, including salary, promotion and tenure, and conditions of work;*

so that faculty can and are encouraged to—

◆ *spend time in schools working with teachers and students;*

◆ *collaborate with schools and teachers in the design of preservice and continuing education programs;*

◆ *offer appropriate graduate courses and programs for experienced teachers of mathematics;*

◆ *provide leadership in conducting and interpreting mathematics education research, particularly school-based research;*

◆ *cooperate with precollege educators to articulate the K–16 mathematics programs;*

◆ *make concerted efforts to recruit and retain teacher candidates of quality and diversity.*

Elaboration

The work of colleges and universities is fundamental to successful mathematics teaching and the education of qualified mathematics teachers. Colleges and universities have major responsibilities for not only the preservice and continuing education of teachers, but also for relevant programs that are responsive to today's and tomorrow's educational needs. Faculty should collaborate with other practicing professionals to design preservice and continuing education programs that reflect the issues of reform and change that must be implemented.

Flexible and alternative methods for continuing education and self-improvement must be instituted to support ongoing learning of mathematics and mathematics education. Graduate programs offered in the late afternoon and in the summer make continued study more accessible for teachers. These learning opportunities need to be particularly attentive to the special needs of adult learners—the practicing mathematics professionals.

Colleges and universities must also work with schools to initiate recruitment and retention efforts to attract quality candidates to the field of mathematics teaching. It is of particular concern that efforts be established to attract, support, and retain minorities and persons with physical disabilities in teaching.

Mathematics educators should participate in school mathematics programs in ways that will keep them current with respect to experience and knowledge about the realities of classroom instruction. The supervision of student teachers requires working closely with cooperating teachers and teacher-candidates for significant time periods. During these times, mathematics educators are not only observing the student teacher, they are gaining understandings about all aspects of the school environment. On another level, mathematics education faculty may become classroom teachers in order to gain first-hand experience with teaching in today's classrooms and to have an environment in which to experiment with teaching and learning. This work is recognized as part of the full teaching and research responsibilities that are required by the college or university and usually is welcomed and facilitated by a local school district. Collegiate faculty use such opportunities as a laboratory for teacher candidates with whom they are working. Collegiate faculty both influence and are influenced by their ongoing interaction with K–12 students, with experienced teachers, and with teacher candidates at their college or university. This involvement with schools is necessary to help articulate mathematics programs K–16.

Colleges and universities have a major responsibility to work with and in schools to develop new knowledge to shape practice. Through basic and applied research in the teaching and learning of mathematics both theoretical and practical knowledge to guide mathematics teaching can be developed. Teachers must be recognized as and encouraged to be partners with college and university faculty in planning, conducting, and interpreting research that impacts on mathematics teaching and learning.

Collegiate faculty should be actively involved in the mathematics and mathematics education communities through participation in a variety of professional organizations. This participation would enable them to share their expertise by making professional contributions to publications and conferences.

In order for collegiate faculty to meet this standard, college and university administrators must establish a reward system including salary, promotion, and tenure that supports and encourages faculty to engage with schools in such work. This calls for a change in the current culture of colleges and universities. Thoughtful analytic work in schools must be recognized as a scholarly activity that is very demanding physically and intellectually and that is critically important to society. In many departments of mathematics, a large percent of the majors are preservice teachers. In the current culture of mathematics departments these preservice teachers and those who work with them are treated as second-class citizens. This must change, as must the attitudes society holds toward mathematics.

STANDARD 4:
PROFESSIONAL ORGANIZATIONS' RESPONSIBILITIES

The leaders of professional organizations need to take an active role in supporting teachers of mathematics by accepting responsibility for—

◆ *promoting and providing professional growth opportunities for those involved in mathematics education;*

◆ *focusing attention of the membership and the broader community on contemporary issues dealing with the teaching and learning of mathematics;*

◆ *promoting activities that recognize the achievements and contributions of exemplary mathematics teachers and programs;*

◆ *initiating political efforts that effect positive change in mathematics education.*

Elaboration

Professional organizations foster a strong sense of community through a variety of strategies, such as written materials, videotapes, journals, and conferences. They provide the vehicles for teachers of mathematics to link with other professionals through the growing use of computer networks and message centers.

Demands for improved mathematics education have been well documented in a variety of reports. Since the mathematics education community should be leading the reform efforts, professional organizations are essential in helping mathematics educators be heard in the vast political community. This necessitates strong and vital organizations that can inform mathematics educators of the current issues, encourage attention to a diversity of points of view about what is important, mobilize efforts to reach consensus on what is needed, present a strong and unified voice for dynamic and thoughtful change, and make this voice heard in the larger corporate, government and policy-making sectors.

Through the work of local, state, provincial, regional, and national organizations, teachers have the support to find avenues to be involved in decision making related to mathematics education. The current work of NCTM on this set of standards and the companion document, *Curriculum and Evaluation Standards for School Mathematics,* as well as the Mathematical Association of America's report, *A Call for Change: Recommendations for the Mathematical Preparation of Teachers* (1991), has involved thousands of mathematics educators in the drafting, reviewing, and revising of the standards. This activity is a prototype for the development of a professional consensus to provide a definitive direction for important aspects of a discipline.

All organizations are challenged to assert their leadership roles and support teachers of mathematics to grow professionally and to achieve greater recognition as respected professionals. The increased status of mathematics educators will influence new candidates to select teaching as a rewarding lifetime career. Above all, teachers must be recognized as the professionals they are.

NEXT STEPS

Teaching is a complex interaction between the teacher, the content being taught, and the students. In order to make change in the teaching and learning of mathematics, each of these—teacher, curriculum and student—must be considered. It is the total environment in which teaching and learning takes place that must be reformed.

In the *Curriculum and Evaluation Standards for School Mathematics* the argument is made that what is needed is a design change strategy This means that new ways of doing things within the system—new roles for teachers and students, new goals, new structures—must be explored to find solutions to persistent problems that result in students failing to become mathematically powerful.

The Role of Standards

The *Curriculum and Evaluation Standards for School Mathematics* is an initial step in this process of design change. It provides guidance for the restructuring of the school mathematics curriculum and for evaluating students and programs. These standards are not a prescription for what must be done at each grade level. They present a vision of what a high-quality mathematics education for students should entail. The *Professional Standards for Teaching Mathematics* provides similar guidance about the kinds of teaching environments, actions, and activities that are needed in order to realize the goals for students that are envisioned in the *Curriculum and Evaluation Standards for School Mathematics*. Teachers, administrators, parents, other educators, and policymakers in government, business, and industry are expected to work collaboratively to reach consensus on *how* their school mathematics programs need to be changed and *what steps* are needed to make that change.

Changing School Mathematics

A dialogue on school reform is taking place on many fronts. In local areas, individual schools, districts, and universities are approaching change in different ways and taking steps in different sequences. What is important is that there be systematic long term commitment to change that heads in an appropriate direction. One of the strengths of the educational system in North America is its diversity. Different combinations of ideas and strategies will provide many ways to achieve the goal of reforming mathematics teaching and learning. In addition to the possible paths discussed in the *Curriculum and Evaluation Standards for School Mathematics*, the following suggestions focus specifically on teaching and the profession of teaching.

Professionalism. At the present time, teaching as a profession does not receive the public support and esteem that it deserves. Teachers often find themselves in positions where decisions that greatly impact their ability to teach are being made by persons who do not have the expertise that teachers have gained through their education and experience. Yet it is the teachers who are held accountable to the public for the mathematics proficiency of their students. A number of efforts are already under way to rethink the roles of teachers as professionals. This movement to raise teaching to a profession with all of the rights and responsibilities entailed is consistent with the vision contained in these Professional Standards for Teaching Mathematics. For example, as the National Board for Professional Teaching Standards moves toward the certification of teachers for differentiated roles in their schools, we expect that it will look to these teaching standards for guidance in developing policies and procedures for the evaluation of mathematics teaching.

Professional mathematics teachers are accountable for teaching mathematics in an intellectually honest and effective way and for making appropriate instructional decisions. Further, they must be an integral part of the ongoing development and regulation of the profession. Mathematics teachers entering teaching should have the support of mentors who are experienced teachers of mathematics. Not only should programs for the professional development of teachers become an established part of school life, but they should be responsive to teachers' needs at all stages of development. As teachers become more experienced and effective, they should be promoted and accepted as leaders in their schools and in the profession as a whole.

Schools have a great deal to gain by supporting teachers' professional development. The teacher is the key to learning in the classroom. Through the individual acts of teachers, the reform of school mathematics will become a reality. Teachers who have the self-esteem and the intrinsic reward that comes from being encouraged to grow in their profession will provide an environment for students in which students see the excitement and usefulness of mathematics. Further, these teachers will provide role models to attract talented students to enter the teaching profession.

Structure of schools. With the growth of professionalism comes the need for a fundamental rethinking of the structure of schools. At the present time teachers are often faced with trying to teach mathematical inquiry in time periods that are entirely inappropriate. Changes, such as meeting classes less often but for a longer period of time, should be explored. Teachers need time to plan, discuss problems with their colleagues, visit another teacher as a peer coach, study, and revise and develop curriculum. Some teachers find it difficult to obtain permission and support to attend professional mathematics meetings. Yet it is through such stimulation that teachers grow and maintain their knowledge about, and enthusiasm for, teaching mathematics and making improvements in classroom instruction. Providing teachers with the support to make instructional decisions is essential. Mathematics teachers often do not have the necessary resources to do their jobs well. They need calculators, computers, software, manipulatives, and other resources to create the kinds of environments for learning that students need and deserve. Giving teachers more responsibility in the budgeting of school resources is also an effective way of improving mathematics instruction. These kinds of changes, and others perhaps not yet conceived, may provide the stimulus needed to effect real change.

We must think creatively and courageously consider changes in the basic structure of schools; try alternatives and carefully study the effects; and create different working models of school structures in which students' mathematical power and teachers' professional growth far exceeds today's models.

Entry into the profession. The current shortage of qualified mathematics teachers and the changing nature of the pool from which teachers come has spurred experimentation with different forms of initial licensure. Induction programs that offer a form of in-school apprenticeship to persons holding undergraduate subject matter degrees are being tried in various places. Other forms of initial licensure for teachers are likely to be tried. These *Professional Standards for Teaching Mathematics* will provide guidance for such induction and licensure programs. Whether teachers enter teaching through four-year, five-year, or induction programs, it is essential that they know the subject matter of mathematics, how students think about mathematics, strategies for teaching mathematics, how to select or create mathematical tasks, and how to create an environment for learning mathematics in which *all* students develop mathematical power.

Teachers of elementary, middle, and high school mathematics need broad and deep knowledge in three fundamental domains: mathematics, mathematics teaching, and students. This knowledge should be appropriate to the level that they will teach or are teaching. However, teachers need to know both where their students have been and where their students are going mathematically. This means that an elementary school teacher needs to have experience with the big ideas of school mathematics at all levels. In addition, the high school teacher needs to understand what significant mathematical ideas are taught prior to high school and how they are represented. Middle school teachers need to be able to connect what they are teaching to the elementary school experiences of students and also to anticipate the growth of mathematical ideas as the students proceed into high school. Therefore, all programs need to be examined carefully to determine whether teachers in the program are being helped to develop a broad range of vision about the mathematics curriculum, student learning, and teaching.

School mathematics as a part of mathematics. Mathematics has changed dynamically in the past twenty-five years, but school mathematics has not reflected these changes. It has become an entity that is not perceived even by teacher education students at colleges and universities as having much to do with "real" mathematics. Part of the responsibility for this schism between school mathematics and the mathematics studied at university rests on the shoulders of university and college mathematicians. Preservice teachers seldom have opportunities to see how the mathematics that they are studying relates to the mathematics of schools. In addition, students at both levels, school and college, are often being taught an outmoded curriculum that has very little to do with what is important in mathematics today. A reexamination of the relationship of school mathematics and mathematics is a necessary part of the needed reform.

Collegiate curriculum. In the *Curriculum and Evaluation Standards for School Mathematics* and the *Professional Standards for Teaching Mathematics* a vision to guide reform of school mathematics curriculum, teaching, and evaluation is articulated. However, if teachers are to change the way they teach, they need to learn significant mathematics in situations where good teaching is modeled. The collegiate community is beginning to examine aspects of the college undergraduate curriculum. We believe that this effort should be expanded to include consideration of the entire undergraduate curriculum and, perhaps even more important, to the models of instruction used in collegiate classrooms. For example, technology and its use in doing, teaching, and learning mathematics is a responsibility of the mathematics community as well as the mathematics education community. Teachers need to learn in technology-rich environments if they are to teach using technology.

Collaboration between schools and universities. As new structures for the professional development of teachers are created, the lines between universities and schools should become blurred. The interaction of university faculty and school faculty as colleagues with different areas of expertise is likely to improve the teaching and learning of mathematics at both the collegiate and school level. Mathematicians have a responsibility to find creative ways to share the excitement of new advances in mathematics both with school teachers and with their students. *The pool of young people who are interested in pursuing professions in mathematics or the sciences is far too small for the needs of society.* From this pool of young people who have an interest in mathematics will come the teachers of tomorrow. Raising the prestige and rewards of teaching is critical to attracting talented and caring young people into teaching.

Of particular concern is the small representation in the scientific pool—and hence in the pool of teachers—of women, ethnic minorities, handicapped, and other underrepresented students. Mathematicians and mathematics educators at all levels have a responsibility to invest time, energy, and their creative talents in finding ways to communicate the excitement and the usefulness of mathematics to young people, and to devising programs that help underrepresented students succeed in their study of mathematics. Beginning at the university level is too late. To have the desired impact we must begin at the elementary school level. Mathematics educators at all levels need to take responsibility and work together to get all students interested in mathematics in elementary school and to help maintain that interest through middle school, high school, and beyond.

As schools and universities strengthen their ongoing communication, mathematics programs can be articulated between elementary school, middle school, high school, and college. Support groups can be established for teachers at all levels who are attempting to implement the standards.

Expectations. One of the curious aspects of our society is that it is socially acceptable to take pride in not being good in mathematics. A phrase often heard by those who teach mathematics is, "Oh, I was never any good at mathematics." Other societies make the assumption that all students can learn mathematics and that learning is a matter of effort. In our society, we are more likely to think that persons are either born with a mathematical mind or they are not. Our expectations have a great deal to do with how we respond to students and consequently to what students believe that they can do. Teachers, counselors, parents, school administrators, and students themselves need to have high expectations that every student can learn mathematics. The two sets of standards challenge us to create learning environments for students and for teachers in which the building of confidence in the learning and doing of mathematics is a primary goal.

Accrediting and certificating agencies. Just as tests are influential in determining the mathematics curriculum, accrediting agencies such as the National Council for the Accreditation of Teacher Education and state or provincial certificating agencies influence collegiate teacher preparation programs. These agencies can be a constraint on needed change or a force for the improvement of teaching. Such agencies can play a positive role for reform as they work with the professional organizations and schools to be sure that their guidelines are in tune with the goals and vision of teaching that the profession espouses.

For example, accrediting agencies can set guidelines that expect schools of education and mathematics departments to be technology-rich environments in which to learn. These standards documents are the consensus of the mathematics and mathematics education communities and can provide guidance to accrediting agencies in determining their criteria for judging mathematics teacher preparation programs.

States and provinces have departments of education that are responsible for the monitoring of school programs, requirements, and offerings. They also are the principal agents for the initial certification of teachers. In both of these roles, state departments of education should review their practices in light of these standards documents. An additional role for departments of education is to assure that professional development opportunities are organized and available to help teachers and schools implement these standards for school mathematics curriculum and teaching.

Networking with other disciplines. An interesting aspect of school reform is that different disciplines are discovering common goals and common strategies for achieving them. For instance, language arts teachers are working on communication, which is a major goal for the mathematics curriculum at all levels. Networking, sharing ideas, learning from each other, and helping to support another discipline in our mathematics classrooms helps everyone succeed. Students can see the commonality of strategies for attacking problems and the help that discussion and argumentation is to the refining of ideas. They also should see this in social studies, science, language arts, and all school subjects. Cross-discipline studies give additional meaning to strategies and concepts and add interest and variety to the learning process. Given the few hours in a school day and the amount of material to be covered, it is in the best interest of teachers in all disciplines to seek common ground and to support each other.

Research. The vision set forth in the Professional Standards for Teaching Mathematics is based on a set of values and beliefs about mathematics teaching and learning that are consistent with current research. However, these standards suggest a research agenda with respect to teacher education and learning to teach. There is much that we need to know that cannot be determined from current practice. We need experimentation and careful research, new structures of schools, new interactions between universities and schools, new teacher education programs, school and university professional development programs, teaching and learning with computing technology and other forms of technology and tools, new forms of instruction in university and school mathematics classes, and other aspects of reform. Researchers are already engaged in accumulating evidence in many of these areas. It is important that these efforts continue and that universities and the mathematics education community value such research. Results of such studies are needed to guide us on the many possible paths to reform in mathematics teaching and learning.

The kind of teaching envisioned in these standards will take time. We need to understand the trade-offs of changing from "covering" a broad set of mathematical topics to more in-depth investigations of, perhaps, fewer mathematical situations. We need to understand better how to meet the mathematical needs of a diverse student population. We need to understand better how small groups and classroom discourse can be used to facilitate students learning to make mathematical judgments and to enhance their mathematical power. The instructional use of manipulatives, calculators, computers, and other tools and technologies for teaching mathematics needs to be continually studied, with a particular concern for the needs of diverse students. As reform proceeds, many other issues will arise that will need careful study. Research is yet another arena in which schools and universities have much to gain by collaborating with each other.

Summary

There are many possible next steps to improving mathematics teaching and learning. If we make a long-term commitment to the standards set forth within this document and in the *Curriculum and Evaluation Standards for School Mathematics*, if we approach the task with the will to persevere, if we are critical of the steps we take, and if we make mid-course corrections, we will make progress toward the goal of developing mathematical power for all students. The picture of mathematics teaching and learning that is presented in these standards is an ambitious one. We will not reach this goal overnight. Such change will take much work and dedication from teachers and many others. However, this effort is essential if we are to improve mathematics learning for our students. We must be impatient enough to take action and patient enough to sustain our efforts until we see results.

We urge you to start by reading both sets of standards. Talk to your colleagues. Discuss these ideas with parents; school and university administrators; and others in government, business, and industry. Collectively and individually set goals for change; establish a plan that will guide change over the next several years. Seek resources to support that plan. Be a part of working to make mathematical power a reality for every student.

REFERENCES

Ball, Deborah L. *Halves, Pieces, and Twoths: Constructing and Using Representational Contexts in Teaching Fractions.* East Lansing, Mich.: National Center for Research on Teacher Education, 1990.

——. "Unlearning to Teach Mathematics." *For the Learning of Mathematics* 8 (1988): 40–48.

Banchoff, Thomas. "Computer Graphics Applications in Geometry: Because the Light Is Better over Here." In *The Merging of Disciplines: New Directions in Pure, Applied, and Computational Mathematics,* edited by R. E. Ewing, K. I. Gross, and C. F. Martin. New York: Springer Verlag, 1986.

Brown, Stephen I., Thomas J. Cooney, and Doug Jones. *Research in Mathematics Teacher Education.* New York: Macmillan Publishing Co., 1985.

Case, Robbie, and Carl Bereiter. "From Behaviorism to Cognitive Development." *Instructional Science* 13 (1984): 141–58.

Clarke, Doug. *The Mathematics Curriculum and Teaching Program.* Canberra, Australia: Curriculum Development Center, 1988.

Cobb, Paul, and Leslie P. Steffe. "The Constructivist Researcher as Teacher and Model Builder." *Journal for Research in Mathematics Education* 14 (1983): 83–94.

Confrey, Jere. "What Constructivism Implies for Teaching." In *Constructivist Views of the Teaching and Learning of Mathematics*, edited by Robert B. Davis, Carolyn A. Maher, and Nel Noddings. Reston, Va.: National Council of Teachers of Mathematics,1990.

Davis, Robert B. *Learning Mathematics: The Cognitive Science Approach to Mathematics Education.* Norwood, N.J.: Ablex, 1984.

Erickson, Frederick, and Gerald Mohatt. "Cultural Organization of Participant Structures in Two Classrooms of Indian Students." In *Doing the Ethnography of Schooling,* edited by G. D. Spindler. New York: Holt, Rinehart & Winston, 1982.

Franklin, Margaret. *Add-ventures for Girls: Building Math Confidence.* Newton, Mass.: WEEA Publishing Center, 1990.

Gilligan, Carol, Nona P. Lyons, and Trudy J. Hanmer. *Making Connections.* Cambridge, Mass.: Harvard University Press, 1990.

Goldin, Gerald A. "Epistemology, Constructivism, and Discovery." In *Constructivist Views of the Teaching and Learning of Mathematics,* edited by Robert B. Davis, Carolyn A. Maher, and Nel Noddings. Reston, Va.: National Council of Teachers of Mathematics, 1990.

Heid, M. Kathleen. "Resequencing Skills and Concepts in Applied Calculus Using the Computer as a Tool." *Journal for Research in Mathematics Education* 19 (1988):3–25.

Hiebert, James, ed. *Conceptual and Procedural Knowledge: The Case of Mathematics.* Hillsdale, N.J.: Lawrence Erlbaum Associates, 1986.

Kulik, Chen-lin C., and James A. Kulik. "Effects of Ability Grouping on Secondary Students: A Meta-Analysis of Evaluation Findings." *American Educational Research Journal* (1982).

Lampert, Magdalene. "How Do Teachers Manage to Teach?" *Harvard Educational Review* 55 (1985):178–94.

——. "Knowing, Doing, and Teaching Mathematics." *Cognition and Instruction* 3 (1986): 305–42.

Lesh, Richard, and Marcia Landau, eds. *Acquisition of Mathematics Concepts and Processes.* New York: Academic Press, 1983.

Macrorie, Ken. *Twenty Teachers.* New York: Oxford University Press, 1984.

Mathematical Association of America. *A Call for Change: Recommendations for the Mathematical Preparation of Teachers.* Washington, D.C.: MAA, 1991.

Mokros, Janice R., Amy S. Weinberg, Lynn T. Goldsmith, and Susan J. Russell. *What's Typical? Children's Ideas about Average.* Cambridge, Mass.: Technical Education Research Center, 1990.

National Council of Teachers of Mathematics. *Curriculum and Evaluation Standards for School Mathematics.* Reston, Va.: The Council, 1989.

"NCTM Policy Statement." *News Bulletin* 27 (November 1990): 3.

_____. "Position Statement on Evaluation of Teacher Performance." Reston, Va.: The Council, 1987.

National Research Council. *Everybody Counts: A Report on the Future of Mathematics Education.* Washington, D.C.: National Academy Press, 1989.

_____. *Reshaping School Mathematics: A Framework for Curriculum.* Washington, D.C.: National Academy Press, 1990.

Noddings, Nel. "Constructivism in mathematics education." In *Constructivist Views of the Teaching and Learning of Mathematics*, edited by Robert B. Davis, Carolyn A. Maher, and Nel Noddings. Reston, Va.: National Council of Teachers of Mathematics, 1990.

Oakes, Jeannie. *Keeping Track: How Schools Structure Inequality.* New Haven, Conn.: Yale University Press, 1985.

Ohio Mathematics Education Leadership Council. *Real Routes: A Handbook for School-based Mathematics Leaders.* Oberlin, Ohio: Oberlin College Press, 1989. [Available from James I. Hassel, Box 433, Berea, OH 44017.]

Resnick, Lauren B. *Education and Learning to Think.* Washington, D.C.: National Academy Press, 1987.

Schoenfeld, Alan H., ed. *Cognitive Science and Mathematics Education.* Hillsdale, N.J.: Lawrence Erlbaum Associates, 1987.

Secada, Walter. "The Challenges of a Changing World for Mathematics Education." In *Teaching and Learning Mathematics in the 1990s,* 1990 Yearbook of the National Council of Teachers of Mathematics, edited by Thomas J. Cooney. Reston, Va.: The Council, 1990.

Slavin, Robert. *Ability Grouping and Student Achievement in Elementary Grades: A Best Evidence Synthesis.* Baltimore: Center for Research on Elementary and Middle Schools, Johns Hopkins University, 1986.

Stenmark, Jean K. *Assessment Alternatives in Mathematics.* Berkeley, Calif.: EQUALS, Lawrence Hall of Science, University of California, 1989.

Weiss, Iris. *Science and Mathematics Education Briefing Book.* Chapel Hill, N.C.: Horizon Research, 1989.

Welch, Wayne. "Science Education in Urbanville: A Case Study." In *Case Studies in Science Education,* edited by R. Stake and J. Easley, p. 6. Urbana, Ill.: University of Illinois, 1978.

FOR MORE COPIES OF THIS AND OTHER IMPORTANT STANDARDS–RELATED PUBLICATIONS...

The future of mathematics education is in your hands. That's why you are receiving a copy of the *Professional Standards for Teaching Mathematics*. With this book and its forerunner, the *Curriculum and Evaluation Standards for School Mathematics*, you have the two most important tools for improving mathematics learning for all students.

For each volume, NCTM has produced a descriptive *Executive Summary*, useful in presenting the visions of the two books to large groups, such as PTAs or seminar participants. These short, 16-page booklets offer a brief overview of the content and goals of the larger volumes.

To order, use the NCTM order form below for NCTM publications only or call (800) 235-7566 or (703) 620-9840. Fax: (703) 476-2970. MasterCard and VISA accepted.

♦ *Professional Standards for Teaching Mathematics* (#439S3)
♦ *Curriculum and Evaluation Standards for School Mathematics* (#396S3)

| 1–9 copies | $25 each |

Special Quantity Discounts:

10–24 copies	20 percent
25–99 copies	30 percent
100–999 copies	40 percent
1000 or more copies	50 percent

♦ *Executive Summary: Professional Standards for Teaching Mathematics* (#480S3)
♦ *Executive Summary: Curriculum and Evaluation Standards for School Mathematics* (#420S3)

Minimum Order: 10 copies

10–49 copies	$1.50 each
50–99 copies	$1.25 each
100 or more copies	$1.00 each

NATIONAL COUNCIL OF TEACHERS OF MATHEMATICS
1906 Association Drive, Reston, VA 22091-1593

ORDER FORM

♦ *Professional Standards for Teaching Mathematics* (#439S3)

		Number	Total
1–9 copies*	$25 each	_____	_____
20% Discount (10–24 copies)	$20.00 each	_____	_____
30% Discount (25–99 copies)	$17.50 each	_____	_____
40% Discount (100–999 copies)	$15.00 each	_____	_____
50% Discount (1000 or more copies)	$12.50 each	_____	_____

♦ *Curriculum and Evaluation Standards for School Mathematics* (#396S3)

1–9 copies*	$25 each	_____	_____
20% Discount (10–24 copies)	$20.00 each	_____	_____
30% Discount (25–99 copies)	$17.50 each	_____	_____
40% Discount (100–999 copies)	$15.00 each	_____	_____
50% Discount (1000 or more copies)	$12.50 each	_____	_____

Copies purchased on the special quantity discount will be shipped to one address only. Copies are nonreturnable and nonrefundable. This special discount policy is not retroactive on previously submitted orders. Shipping and handling charges will be added to all billed orders.

*A 20% discount is available for individual members.

♦ *Executive Summary: Professional Standards for Teaching Mathematics* (#480S3)
♦ *Executive Summary: Curriculum and Evaluation Standards for School Mathematics* (#420S3)

Minimum Order: 10 copies		Number	Total
10–49 copies	$1.50 each	_____	_____
50–99 copies	$1.25 each	_____	_____
100 or more copies	$1.00 each	_____	_____

Total Enclosed (Va. residents add 4.5% sales tax) | $ _____ |

Please charge my ☐ MasterCard or ☐ VISA.

CARD # _____ EXP. DATE _____

SIGNATURE _____

NAME _____

ADDRESS _____

CITY _____

STATE/PROVINCE _____ POSTAL CODE _____

Mail to: National Council of Teachers of Mathematics, 1906 Association Drive, Reston, VA 22091-1593

COUNTING ON YOU: Supporting Standards for Mathematics Teaching, a publication of the Mathematical Sciences Education Board, is available by writing to:

National Academy Press
2101 Constitution Avenue, NW, Box 285
Washington, DC 20055

or calling: (800) 624-6242
(202) 334-3313 in the Washington, D.C., area.

Single copies: $2.95
Orders of ten copies or more: $2 each

Make checks payable to: National Academy Press.
Payment must be enclosed with order.

NAME _____

ADDRESS _____

CITY _____

STATE/PROVINCE _____ POSTAL CODE _____

A CALL FOR CHANGE: RECOMMENDATIONS FOR THE MATHEMATICAL PREPARATION OF TEACHERS OF MATHEMATICS is available from:

Mathematical Association of America
1529 Eighteenth Street, NW
Washington, DC 20036

Please send me _____ copies (Catalog Number CFC) at $7 each for a total of $_____. Make check or money order payable to MAA. (Payment for foreign orders must be made in U.S. funds.) Orders under $10 must be prepaid. Shipping and handling charges will be added to all billed orders and all foreign orders. 20% discount available to bookstores.

NAME _____

ADDRESS _____

CITY _____

STATE/PROVINCE _____ POSTAL CODE _____

Please charge my ☐ MasterCard or ☐ VISA.

CARD # _____ EXP. DATE _____

SIGNATURE _____